Mathematical Physics Studies

The series publishes original research monographs on contemporary mathematical physics. The focus is on important recent developments at the interface of Mathematics, and Mathematical and Theoretical Physics. These will include, but are not restricted to: application of algebraic geometry, D-modules and symplectic geometry, category theory, number theory, low-dimensional topology, mirror symmetry, string theory, quantum field theory, noncommutative geometry, operator algebras, functional analysis, spectral theory, and probability theory.

More information about this series at http://www.springer.com/series/6316

Akinori Tanaka • Akio Tomiya • Koji Hashimoto

Deep Learning and Physics

Akinori Tanaka
iTHEMS
RIKEN
Wako, Saitama, Japan

Akio Tomiya
Radiation Lab
RIKEN
Wako, Saitama, Japan

Koji Hashimoto
Department of Physics
Osaka University
Toyonaka, Osaka, Japan

ISSN 0921-3767 ISSN 2352-3905 (electronic)
Mathematical Physics Studies
ISBN 978-981-33-6110-2 ISBN 978-981-33-6108-9 (eBook)
https://doi.org/10.1007/978-981-33-6108-9

This Springer imprint is published by the registered company Springer Nature Singapore Pte Ltd.
The registered company address is: 152 Beach Road, #21-01/04 Gateway East, Singapore 189721, Singapore

To readers traveling beyond the boundaries of knowledge

Preface

What is deep learning for those who want to study physics? Is it completely different from physics? Or is it really similar?

In recent years, machine learning, including deep learning, has begun to be used in various physics studies. Why is that? Is knowing physics useful in machine learning? Conversely, is knowing machine learning useful in physics?

This book is devoted to answers of these questions. Starting with basic ideas of physics, neural networks are derived "naturally." And you can learn the concepts of deep learning through the words of physics.

In fact, the foundation of machine learning can be attributed to physical concepts. Hamiltonians that determine physical systems characterize various machine learning structures. Statistical physics given by Hamiltonians defines machine learning by neural networks. Furthermore, solving inverse problems in physics through machine learning and generalization essentially provides progress and even revolutions in physics. For these reasons, in recent years, interdisciplinary research in machine learning and physics has been expanding dramatically.

This book is written for anyone who wants to know, learn, and apply the relationship between deep learning/machine learning and physics.[1] All you need to read this book is just the basic concepts in physics: energy and Hamiltonians.[2] The concepts of statistical mechanics and the bracket notation of quantum mechanics, which are introduced in columns, are used to explain deep learning frameworks.

This book is divided into two parts. The first part concerns understanding the machine learning method from the perspective of physics, and the second part

[1] If the reader has learnt physics and then reads a general machine learning textbook, the reader may feel uncomfortable with how various concepts are introduced suddenly and empirically without explanation. This book is motivated by such gaps. On the other hand, code implementation is not covered in this book, because it is library dependent. The reader can try to implement a machine with some favorite library.

[2] It is assumed that readers learned the basics of physics: around second or third grade of undergraduate courses in physics. We will explain everything based on Hamiltonians, but it is not necessary to have knowledge on analytical mechanics.

explores the relationship between various problems in physics and the method of machine learning. That is, the first part is written as a textbook, but the second part is not a text book, it is a collection of recent advanced topics (and so, the chapters in the second part can be read almost independently). Please refer to the reader's guide at the beginning of each section.

A renowned physicist, Ryoyu Uchiyama, said in the preface to his book, Theory of Relativity [1], "If you read this book and do not understand it, you should give up learning the theory of relativity." On the other hand, the field of "machine learning × physics," which is the subject of this book, is not a discipline established over a long history like the theory of relativity, but a field of research that is still undergoing great progress. So, rather, our message is this: "If you read this book and do not understand it, it will be a seed for development in the future."

One of our favorites is the words by S. Weinberg, a particle physicist [2]:

"My advice is to go for the messes – that's where the action is."

It is up to the readers to make their own way. We hope this book helps the readers learn and study.

Tokyo, Japan
March 2019

Akinori Tanaka
Akio Tomiya
Koji Hashimoto

The authors

Acknowledgments

This book was written based on discussions about physics and machine learning with many people around us. Particularly essential were the discussions at the symposia "Deep Learning and Physics" and "Deep Learning and Physics 2018" held at Osaka University, as well as at research meetings and seminars around the world. We would like to take this opportunity to thank all those involved. We thank Tetsuya Akutagawa, Naoyuki Sumimoto, Norihito Shirai, Yuki Nagai, Yuta Kikuchi, Shuji Kitora, and Sayaka Sone for reading and commenting on the Japanese draft. Akinori Tanaka gratefully acknowledges support from the RIKEN Center for Advanced Intelligence Project and RIKEN Interdisciplinary Theoretical and Mathematical Sciences Program (iTHEMS). Akio Tomiya thanks Prof. Heng-Tong Ding at Central China Normal University. If he had not hired Akio as a postdoc and given Akio an environment to freely study issues which are different from his profession, the papers that sparked this book would not have been possible. Akio would like to thank Prof. Makoto Matsumoto at Hiroshima University for pointing out the relationship between true random numbers and the Kolmogorov complexity. Koji Hashimoto thanks his wife, Haruko Hashimoto, who understands writing and research well. We all thank Haruko Hashimoto for her support on English translation and Mr. Atsushi Keiyama, an editor at Kodansha Scientific, for his effort on the publication of the Japanese version of this book (published in June 2019). We are grateful to Prof. Hirosi Ooguri and Prof. Masatoshi Imada who kindly supported the publication of this book at Springer, and also to Mr. Nakamura for his editorial work on this book. Finally, we would like to thank the researchers from all over the world for discussing physics and machine learning at various locations.

Contents

Chapter 1
Forewords: Machine Learning and Physics

Abstract What is the relationship between machine learning and physics? First let us start by experiencing why machine learning and physics can be related. There is a concept that bridges between physics and machine learning: that is information. Physics and information theory have been mutually involved for a long time. Also, machine learning is based on information theory. Learning is about passing information and recreating relationships between information, and finding information spontaneously. Therefore, in machine learning, it is necessary to use information theory that flexibly deal with the amount of information, and as a result, machine learning is closely related to the system of information theory. This chapter explores the relationship between physics, information theory, and machine learning, the core concepts in this book.

What is the relationship between machine learning and physics? We'll take a closer look at that in this book, but first let us start by experiencing why machine learning and physics can be related. There is a concept that bridges between physics and machine learning: that is information.

Physics and information theory have been mutually involved for a long time, and the relationship is still widely and deeply developed. Also, machine learning is based on information theory. Learning is about passing information and recreating relationships between information, and finding information spontaneously. Therefore, in machine learning and deep learning, it is necessary to use information theory that flexibly deals with the amount of information, and as a result, machine learning is closely related to the system of information theory.

As the reader can imagine from these things, machine learning and physics should have some big relationship with "information" as an intermediate medium. One of the goals of this book is to clarify this firm bridge. Figure 1.1 shows a conceptual diagram.

This chapter explores the relationship between physics, information theory, and machine learning, the core concepts in this book. Let us explain how the titles of this book, "Deep Learning" and "Physics" are related.

A. Tanaka et al., *Deep Learning and Physics*, Mathematical Physics Studies,
https://doi.org/10.1007/978-981-33-6108-9_1

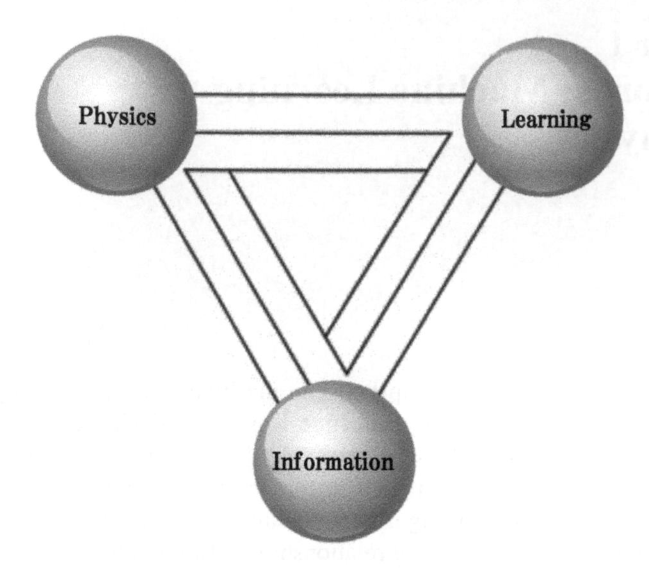

Fig. 1.1 Physics, machine learning, and information. Do they form a triangle?

1.1 Introduction to Information Theory

Quantifying information

This chapter explores the relationship between physics and machine learning using information theory as a bridge. For that, we need to define exactly what "information" is. What is "information" in the first place? First, read the following two sentences [3]:

- Humans teach gorillas how to add numbers. (1.1)
- Gorillas teach humans how to add numbers. (1.2)

Which of these two sentences can be said to have more **information**? (1.1) is not surprising, if true, because it is possible, regardless of whether the gorilla understands it or not. You may have actually heard such news. On the other hand, if (1.2) is true, it is quite surprising. Then one will change one's mind and remember that there could be a gorilla which can teach humans. This may be said to be increasing the amount of information. In other words [4],

$$\text{Amount of information} = \text{Extent of surprise.} \tag{1.3}$$

Let us go ahead with this policy anyway. The greater the surprise, the less likely it is to happen. In addition, if we try to make "information increase by addition", with

P(event) the probability that an event will occur, we have[1]

$$\text{Amount of information of event A} = -\log P(\text{event A}). \tag{1.4}$$

When the probability is low, the amount of information is large.

Average amount of information
Let us further assume that various events $A_1, A_2, \ldots, A_W$ occur with probabilities $p_1, p_2, \ldots, p_W$, respectively. At this time, the amount of information of each event is $-\log p_i$, and its expectation value

$$S_{\text{information}} = -\sum_{i=1}^{W} p_i \log p_i. \tag{1.5}$$

is called **information entropy**.[2] Let us take a concrete example of what information entropy represents. Suppose there are W boxes, and let p_i be the probability that the ith box contains a treasure. Of course, we want to predict which box contains the treasure as accurately as possible, but the predictability depends on the value of p_i. For example, if we know that the treasure is always in the first box, it is easy to predict. The value of the information entropy for this case is zero:

$$p_i = \begin{cases} 1 \ (i = 1) \\ 0 \ (\text{other than that}) \end{cases} \qquad S_{\text{information}} = 0. \tag{1.6}$$

On the other hand, if the probability is completely random,

$$p_i = \frac{1}{W} \qquad S_{\text{information}} = \log W. \tag{1.7}$$

For this case, even if we know the probability, it is difficult to predict because we do not know which box it is in. This time, the information entropy has a large value, $\log W$. In other words, the more difficult it is to predict, the greater the information entropy. Therefore, the relation to the commonly referred to **"information"** is as follows:

$$\bullet \begin{cases} \text{Little "information"} \Leftrightarrow \text{difficult to predict} \Leftrightarrow \text{large information entropy} \\ \text{A lot of "information"} \Leftrightarrow \text{easy to predict} \Leftrightarrow \text{small information entropy} \end{cases} \tag{1.8}$$

[1]In this book, as in many physics textbooks, the base of the logarithm is taken to be that of the natural log, $e = 2.718\cdots$.

[2]This quantity was first introduced in the context of information in the monumental paper by C. Shannon on mathematics in communication [5]. He calls this quantity entropy, so physicists can easily understand this concept.

The "information" here is the information we already have, and the amount of information (1.4) is the information obtained from the event.

1.2 Physics and Information Theory

There are many important concepts in physics, but no physicist would oppose that one of the most important is entropy. Entropy is an essential concept in the development of thermodynamics, and is expressed as $S = k_B \log W$, where W is the number of microscopic states, in statistical mechanics. The proportionality coefficient k_B is the Boltzmann constant and can be set to 1 if an appropriate temperature unit is used. The entropy of the system in this case is

$$S = \log W . \tag{1.9}$$

In physics, the relationship with information theory is revealed through this entropy: That is because $S_{\text{information}}$ of (1.7) is exactly the same formula as (1.9). In this way, various discussions on physical systems dealing with multiple degrees of freedom, such as thermodynamics and statistical mechanics, have a companion information-theoretic interpretation.

By the way, most of the interest in research in modern physics (e.g., particle physics and condensed matter physics) is in many-body systems. This suggests that information theory plays an important role at the forefront of modern physics research. Here are two recent studies.

Black hole information loss problem

Research by J. Bekenstein and S. Hawking [6, 7] shows theoretically that black holes have entropy and radiate their mass outward as heat. To explain the intrinsic problem hidden in this radiation property of black holes, the following example is instructive: Assume that there is a spring, and fix it while stretched. If this is thrown into a black hole, the black hole will grow larger by the stored energy, and then emit that energy as heat with the aforementioned thermal radiation. But here is the problem. Work can be freely extracted from the energy of the spring before it is thrown, but the second law of thermodynamics limits the energy efficiency that can be extracted from the thermal radiation. In other words, even if there is only a single state before the throwing (that is, the entropy is zero), the entropy increases by the randomness after thermal radiation. An increase in entropy means that information has been lost (see (1.8)). This is known as the information loss problem and is one of the most important issues in modern physics for which no definitive solution has yet been obtained.[3]

[3]For this, take a look at the column in Chap. 12.

Maxwell's demon

Maxwell's demon is a virtual devil that appears in a thought experiment and breaks the second law of thermodynamics, at least superficially; it was introduced by James Clerk Maxwell. Assume a box contains a gas at temperature T; a partition plate with a small hole is inserted in the middle of this box. The hole is small enough to allow one gas molecule to pass through. There is also a switch next to the hole, which can be pressed to close or open the hole. According to statistical mechanics, in a gas with a gas molecular mass m and temperature T, molecules with speed v exist with probability proportional to $e^{-\frac{mv^2}{2k_B T}}$. This means that gas molecules of various speeds are flying: there are some fast molecules, and some slow molecules. Assuming that a small devil is sitting near the hole in the partition plate, the devil lets "only the fast molecules coming from the right go through the hole to the left, only the slow molecules from the left go through the hole to the right." As a result, relatively slow molecules remain on the right, and fast-moving molecules gather on the left. That is, if the right temperature is T_R and the left temperature is T_L, it means that $T_R < T_L$. Using the ideal gas equation of state, $p_R < p_L$, so the force $F = p_L - p_R$ acts in the direction of the room on the right. If we allow the partition to slide and attach some string to it, then this F can do some work. There should be something strange in this story, because it can convert heat to work endlessly, which means that we have created a perpetual motion machine of the second kind. In recent years, information theory has been shown to be an effective way to fill this gap [8].

In this way, relations with information theory have been more and more prominent in various aspects of theoretical physics. Wheeler, famous for his work on gravity theory, even says "it from bit" (physical existence come from information) [9]. In recent years, not only information theory based on ordinary probability theory, but also research in a field called quantum information theory based on quantum mechanics with interfering probabilities has been actively performed, and various developments are reported daily. This is not covered in this book, but interested readers may want to read [10] and other books.

1.3 Machine Learning and Information Theory

One way to mathematically formulate machine learning methods is based on probability theory. In fact, this book follows that format. One of the nice things about this method is that we can still use various concepts of information theory, including entropy. The purpose of machine learning is "to predict the future unknown from some experience," and when formulating this mathematically, as in (1.8), it is necessary to deal with a quantity measuring the degree of difficulty in predicting things. However, the "predictability" described in (1.8) is based on the assumption that we know the probability p_i of the occurrence of the event. Even in machine learning, it is assumed that there exists p_i behind the phenomenon, while its value is not known. The following example illustrates that even in such cases, it is still very important to consider a concept similar to entropy.

Relative entropy and Sanov's theorem
Here, let us briefly look at a typical method of machine learning which we study in this book.[4] As before, let us assume that the possible events are $A_1, A_2, \ldots, A_W$, and that they occur with probabilities $p_1, p_2, \ldots, p_W$, respectively. If we can actually know the value of p_i, we will be able to predict the future to some extent with the accuracy at the level of the information entropy. However, in many cases, p_i is not known and instead we only know "information" about how many times A_i has actually occurred,

$$\bullet \begin{cases} A_1 : \#_1 \text{ times}, A_2 : \#_2 \text{ times}, \ldots, A_W : \#_W \text{ times}, \\ \#(= \sum_{i=1}^{W} \#_i) \text{ times in total.} \end{cases} \tag{1.10}$$

Here, # (the number sign) is an appropriate positive integer, indicating the number of times. Just as physics experiments cannot observe the theoretical equations themselves, we cannot directly observe p_i here. Therefore, consider creating an expected probability q_i that is as close as possible to p_i and regard the problem of determining a "good" q_i here as machine learning. How should we determine the value of q_i from the "information" (1.10) alone? One thing we can do is to evaluate

- Probability of obtaining information (1.10) assuming q_i is the true probability. (1.11)

If we can calculate this, we need to determine q_i that makes the probability (1.11) as large as possible (close to 1). This idea is called the **maximum likelihood estimation**. First, assume that each A_i occurs with probability q_i,

$$p(\text{probability of } A_i \text{ occurring } \#_i \text{ times}) = q_i^{\#_i}. \tag{1.12}$$

Also, in this setup, we assume that the A_is can occur in any order. For example, $[A_1, A_1, A_2]$ and $[A_2, A_1, A_1]$ are counted as the same, and the number of such combinations should be accounted for in the probability calculation. This is the multinomial coefficient

$$\binom{\#}{\#_1, \#_2, \ldots, \#_W} = \frac{\#!}{\#_1!\#_2! \cdots \#_W!}. \tag{1.13}$$

Then we can write the probability as the product of these,

$$(1.11) = q_1^{\#_1} q_2^{\#_2} \ldots q_W^{\#_W} \frac{\#!}{\#_1!\#_2! \ldots \#_W!}. \tag{1.14}$$

[4] This argument is written with the help of an introductory lecture note on information theory [11].

Then, we should look for q_i that makes this value as large as possible. In machine learning, q_i is varied to actually increase the amount equivalent to (1.14) as much as possible.[5]

By the way, if the number of data is large ($\# \approx \infty$), by the law of large numbers,[6] we have

$$\frac{\#_i}{\#} \approx p_i \quad \Leftrightarrow \quad \#_i \approx \# \cdot p_i \,. \tag{1.17}$$

Here, $\#_i$ must also be a large value, so according to Stirling's formula (which is familiar in physics), we find

$$\#_i! \approx \#_i^{\#_i} \,. \tag{1.18}$$

By substituting (1.17) and (1.18) into (1.14), we can get an interesting quantity:

$$\begin{aligned}
(1.14) &\approx q_1^{\#\cdot p_1} q_2^{\#\cdot p_2} \dots q_W^{\#\cdot p_W} \frac{\#!}{(\# \cdot p_1)!(\# \cdot p_2)! \dots (\# \cdot p_W)!} \\
&\approx q_1^{\#\cdot p_1} q_2^{\#\cdot p_2} \dots q_W^{\#\cdot p_W} \frac{\#^{\#}}{(\# \cdot p_1)^{\#\cdot p_1}(\# \cdot p_2)^{\#\cdot p_2} \dots (\# \cdot p_W)^{\#\cdot p_W}} \\
&= q_1^{\#\cdot p_1} q_2^{\#\cdot p_2} \dots q_W^{\#\cdot p_W} \frac{1}{p_1^{\#\cdot p_1} p_2^{\#\cdot p_2} \dots p_W^{\#\cdot p_W}} \\
&= \exp\Big[-\# \sum_{i=1}^{W} p_i \log \frac{p_i}{q_i} \Big] .
\end{aligned} \tag{1.19}$$

The goal is to make this probability as close to 1 as possible, which means to bring $\sum_{i=1}^{W} p_i \log \frac{p_i}{q_i}$ close to zero. This quantity is called relative entropy, and is known to be zero only when $p_i = q_i$. Therefore, bringing the original goal (1.11) as close

---

[5]This problem can be solved by the Lagrange multiplier method. Define a Lagrangian

$$L(q_i, \lambda) = \log[\text{Eq. } (1.14)] + \lambda \left(1 - \sum_{i=1}^{W} q_i \right) . \tag{1.15}$$

If we look for q_i, λ which extremize this L, then we find

$$q_i = \frac{\#_i}{\#}, \quad \lambda = \#. \tag{1.16}$$

The reader can see that this fits with the intuition from the law of large numbers (1.17).

[6]The law of large numbers will be explained in Chap. 5.

to 1 as possible corresponds to reducing the relative entropy. We mean,

Relative entropy = Amount to measure how close the prediction q_i is to the truth p_i .

The properties of the relative entropy will be explained later in this book. The relative entropy is called **Kullback-Leibler divergence** in information theory, and is important in machine learning as can be seen here, as well as having many mathematically interesting properties. The fact that (1.11) is approximately the same as the Kullback-Leibler divergence is one of the consequences from large deviation theory called Sanov's (I. N. Sanov) theorem [12, 13].

1.4 Machine Learning and Physics

So far, we have briefly described the relationship between physics and information theory, and the relationship between machine learning and information theory. Then, there may be a connection between machine learning and physics in some sense. The aforementioned Fig. 1.1 shows the concept: physics and information are connected, and information and machine learning are connected. Then, how can physics and machine learning be connected?

A thought experiment
Suppose we have a fairy here. A button and a clock are placed in front of the fairy, and every minute, the fairy chooses to press the button or not. If the fairy presses the button, the machine outputs 1. If the fairy does not press the button, the output is 0. The result of "a special fairy" is the following sequence:

$$\{1, 0, 0, 0, 0, 0, 0, 0, 0, 1, 1, 0, 0, 1, 1, \dots\}. \tag{1.20}$$

The following figure shows the result of this monotonous job for about 4 hours, displayed in a matrix of 15 × 15:

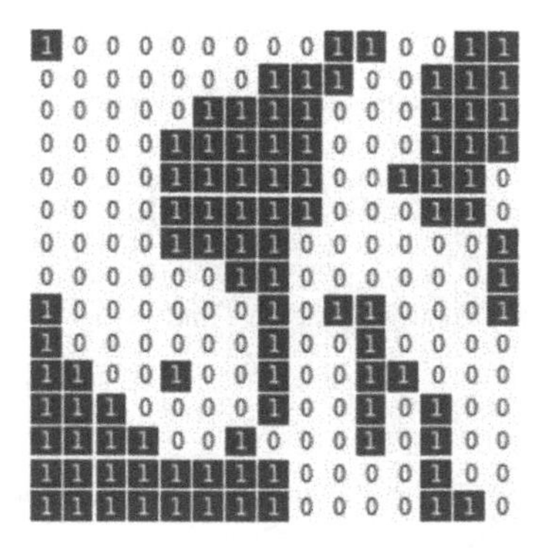

We let the fairy to continue its work all night for five days. The result is:

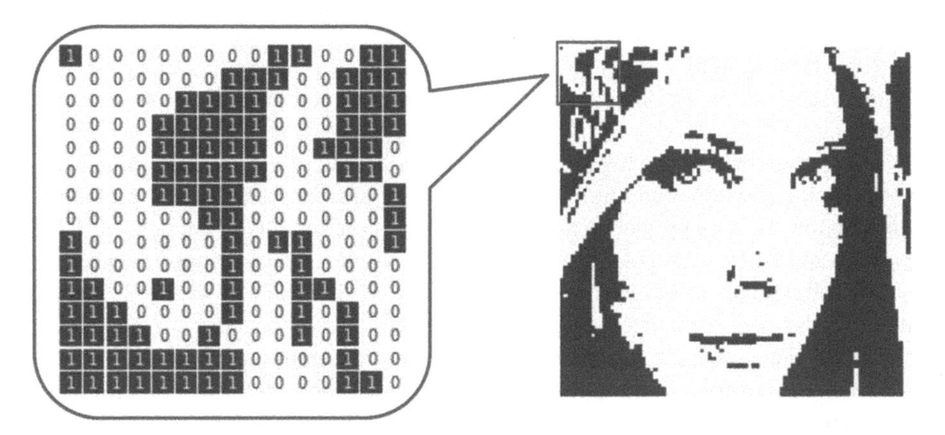

It looks like a person's face.[7] The "special fairy" had the goal of drawing this picture by pressing/not-pressing buttons. This may seem like a trivial task, but here we explain a bit more physically that it is not.

First, consider the random arrangement of 0 and 1 as the initial state. With this we do not lose the generality of the problem. The fairy decides whether to replace 0/1 with 1/0 or leave it as it is at the time (x, y) (coordinates correspond to the time), by operating the button. The fairy checks the state of the single bit at time (x, y) and decides whether or not to invert it. Yes, the readers now know that the identity of this "special fairy" is **Maxwell's demon.**[8]

From this point of view, the act of "drawing a picture" is to draw a meaningful combination (low entropy) from among the myriad of possibilities (high entropy) that can exist on the canvas. In other words, it is to make the information increase in some way. You may have heard news that artificial intelligence has painted pictures. Without fear of misunderstanding, we can say that it means a successful creation of a Maxwell's demon, in the current context.

The story of Maxwell's demon above was a thought experiment that suggested that machine learning and physics might be related, but in fact, can the subject of this book, machine learning and physics, be connected by a thick pipe?

In fact, even in the field of machine learning, there are many problems that often require "physical sense" and there are many research results inspired by it. For example, in the early days of deep learning, a model (Boltzmann machine) that used a stochastic neural network was used, and this model was inspired by statistical

[7]This is a binary image of the sample data "Lenna image" [14]. It is used as a standard in research fields such as image compression.

[8]This analogy is a slightly modified version of the story in [15]. It would be interesting to discuss the modern physics solution of the Maxwell's demon problem in the context here, namely, where does the entropy reduction used to learn some meaningful probability distribution in machine learning come from?

mechanics, especially the Ising model that deals with spin degrees of freedom. A neural network is an artificial network that simulates the connection of neurons in the brain, and the Hopfield model, which provides a mechanism of storing memory as a physical system, can be said to be a type of Ising model. Focusing on these points, in this book we describe various settings of neural networks from statistical mechanics.

In Boltzmann machine learning and Bayesian statistics, it is necessary to generate (or sample) (pseudo-)random numbers that follow some complicated probability distribution. It is a popular method in numerical calculations in condensed matter physics and elementary particle physics, so there is something in common here.[9]

Sampling takes too long for complex models, so in recent years the Boltzmann machine has been replaced by neural networks and is not attracting attention. However, the academic significance may yet be recealed, so it would be a good idea not only to follow the latest topics of deep learning, but also to return to the beginning and rethink the Boltzmann machine.

Today, deep learning is simply a collection of various techniques that have been found to work "empirically" when applying deep neural networks to various machine learning schemes. However, the mind that senses any technique to work "empirically" has something close to the so-called "physical sense".[10] For example, in recent years, it has been known that a model called ResNet, which includes a "bypass" in a neural network, has a high capability. ResNet learns the "residual" of the deep neural network, rather than learning directly the "features" that would have been acquired by the ordinary deep neural networks. The "features" are the quantities important for performing the desired task. In the ResNet, the residuals are accumulated to express the features. This is expressed as a relationship "differential = residual" and "integral = feature", and is reminiscent of the equation of motion and its solution.

In this sense, machine learning and physics seem to have "a certain" relationship. To actually try the translation, we need to create a corresponding dictionary such as Table 1.1. Returning to Fig. 1.1, combined with the many fragmentary relationships listed above, machine learning and deep learning methodologies can be viewed and constructed from a physical perspective.

With this in mind, this book is aimed at:

1. Understanding machine learning methods from a physics perspective

[9]In fact, it is not unusual that people majoring in computational physics switch to the world of data science.

[10]As an example, most quantum field theories have not been mathematically justified, but are empirically consistent with experimental results. They are formulated by a physical sense. Calculated results of quantum electrodynamics, which is a branch of the quantum field theories, contain a number of operations that are not mathematically justified, but they are more than 10 orders of magnitude consistent with experimental measurements. Of course, the performed operations are consistent in the sense of theoretical physicists, but they have not been justified in a mathematically rigorous way.

Table 1.1 Terms and notations used in machine learning related fields and their counterparts in physics

Machine learning	Physics
Expection value $\mathbb{E}_\theta[\bullet]$	Expection value $\langle \bullet \rangle_J$
Training parameters W, θ	Coupling constant (external field) J
Energy function E	Hamiltonian H

2. Finding the link between physics problems and machine learning methods

In this book, the first half is "Introduction to Machine Learning/Deep Learning from the Perspective of Physics", and the second half is "Application to Physics". Although there are many topics which are not introduced in this book, we have tried to include references as much as possible, so readers who want to know more about them should consult the literature or search by keywords.

In addition, the final chapter of this book describes the background of the writing, the motivation for the research, and a message for those who are studying machine learning, from three different perspectives of the authors. As the readers can find there, the authors have different interconnecting opinions, and in what sense machine learning and physics are still related is a mystery that researchers do not completely agree with. In other words, this also means "an unexplored land where something may exist." With the readers who study machine learning from now, let us step into this mysterious land, through this book.

Part I
Physical View of Deep Learning

Let us get into Part I of this book, in which we study and understand machine learning from a physics perspective. In this art, we will use physics language to see how neural networks "emerge." In general textbooks, neural networks used in machine learning and deep learning are introduced as imitating the functions of brain neural networks and neurons, and various neural network structures have been improved for the purpose of learning and application. No particular physics perspective is used there. As a result, many readers may suspect that, even if neural networks have a physics perspective, they can only be interpreted as retrospective interpretations of individual concepts. However, in Part I, from a physics point of view, we will focus on how neural networks and their accompanying concepts emerge naturally, and mainly look at the "derivation" of neural networks.

Incidentally, although we cannot say for sure, we imagine that the "derivation" here seems to be something that was actually done implicitly in the minds of researchers. This is our feeling after having read the papers written at the dawn of deep learning research when the research subject was shifting from Boltzmann machines to neural nets.

Of course, not all machine learning concepts are physically derived. We will explain basic concepts such as what is machine learning, in physics language. If you have learned a bit of physics, you can naturally feel the answer to the question of how to perform and optimize the machine learning.

In addition, the concepts described below do not cover all of the concepts in machine learning in an exhaustive manner, but focus only on the basic concepts necessary to understand what machine learning is. However, understanding these in the language of physics will help to physically understand machine learning that is currently applied in various ways, and will also be useful for research related to physics and machine learning described in Part II. Now, we shall give each chapter a brief introduction as a destination sign for readers.

Chapter 2: Introduction to Machine Learning First, we learn the general theory of machine learning. We shall take a look at examples of what learning is, what is the meaning of "machines learned," and what relative entropy is. We will learn how to handle data in probability theory, and describe "generalization" and its importance in learning.

Chapter 3: Basics of Neural Networks Next, in this chapter, we derive neural networks from the viewpoint of physical models. A neural network is a nonlinear function that maps an input to an output, and giving the network is equivalent to giving a function called an error function in the case of supervised learning. By considering the output as dynamical degrees of freedom and the input as an external field, various neural networks and their deepened versions are born from simple Hamiltonians.[1] Training (learning) is a procedure for reducing the value of the error function, and we will learn the specific method of backpropagation using the bra-ket notation popular in quantum mechanics. And we will look at how the "universal approximation theorem" works, which is why neural networks can express connections between various types of data.

Chapter 4: Advanced Neural Networks In this chapter, we explain the structure of the two types of neural networks that have been the mainstays of deep learning in recent years, following the words of physics in the previous chapter. A convolutional neural network has a structure that emphasizes the spatial proximity in input data. Also, recurrent neural networks have a structure to learn input data in time series. You will learn how to provide a network structure that respects the characteristics of data.

Chapter 5: Sampling In the situation where the training is performed, it is assumed that the input data is given by a probability distribution. It is often necessary to calculate the expectation value of the function of various input values given by the probability distribution. In this chapter, we will look at the method and the necessity of "sampling," which is the method of performing the calculation of the expectation value. The frequently used concepts in statistical mechanics, such as the law of large numbers, the central limit theorem, the Markov chain Monte Carlo method, the principle of detailed balance, the Metropolis method, and the heatbath method, are also used in machine learning. Familiarity with common concepts in physics and machine learning can lead to an understanding of both.

Chapter 6: Unsupervised Deep Learning At the end of Part I we will explain Boltzmann machines and generative adversarial networks (GANs). Both models are not the "find an answer" network given in Chap. 3, but rather the network itself giving the probability distribution of the input. Boltzmann machines have historically been the cornerstone of neural networks and are given by the Hamiltonian statistical mechanics of multi-particle spin systems. It is an important bridge between machine learning and physics. Generative adversarial networks are also one of the important topics in deep learning in recent years, and we try to provide an explanation of them from a physical point of view.

---

[1] For the basic concepts of physics used in this book, statistical mechanics and quantum mechanics, see the columns at the end of chapters.

In Part I, the key point is how to model the mathematics of learning as a physical model and, as a result, obtain the fact that the neural network is derived physically and naturally. These will be useful for various occasions: for further learning about various constructions of machine learning architecture, for understanding new learning methods that will show up in the future, for applying physics to the research of machine learning, and for applying machine learning to the study of physics.

Chapter 2
Introduction to Machine Learning

Abstract In this chapter, we learn the general theory of machine learning. We shall take a look at examples of what learning is, what is the meaning of "machines learned," and what relative entropy is. We will learn how to handle data in probability theory, and describe "generalization" and its importance in learning.

2.1 The Purpose of Machine Learning

Deep learning is a branch of what is called **machine learning**. The old "definition" of machine learning by computer scientist Arthur Samuel [16] is "Field of study that gives computers the ability to learn without being explicitly programmed." In more modern terms, "From experience alone, let the machine automatically gain the ability to extract the structure behind it and apply it to unknown situations."[1]

There are various methods depending on how the experience is given to the machine, so here let us assume a situation in which the experience is already stored in a database. In this context, there are two types of data to consider:

- Supervised data$\{(\mathbf{x}[i], \mathbf{d}[i])\}_{i=1,2,\ldots,\#}$
- Unsupervised data$\{(\mathbf{x}[i], \ - \ )\}_{i=1,2,\ldots,\#}$

Here $\mathbf{x}[i]$ denotes the i-th data value (generally vector valued), and $\mathbf{d}[i]$ denotes **teaching signal**. The mark "$-$" indicates that no teaching signal is given, and in the case of unsupervised data, only $\{\mathbf{x}[i]\}$ is actually provided. # indicates the number of data. See, for example, Fig. 2.1.

[1] Richard Feynman said [17], "We can imagine that this complicated array of moving things which consists "the world" is something like a great chess game being played by the gods, and we are observers of the game. We do not know what the rules of the game are; all we are allowed to do is to watch the playing. Of course, if we watch long enough, we may eventually catch on to a few of the rules." It is a good parable that captures the essence of the inverse problem of guessing rules and structures. The machine learning, compared with this example, would mean that the observer is a machine instead of a human.

A. Tanaka et al., *Deep Learning and Physics*, Mathematical Physics Studies,
https://doi.org/10.1007/978-981-33-6108-9_2

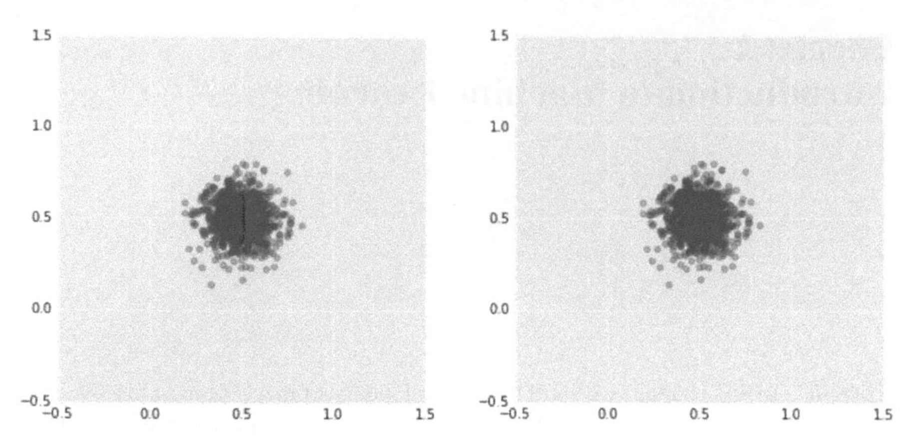

Fig. 2.1 Left: Example of supervised data ($d[i] = 0$ corresponds to red, $d[i] = 1$ corresponds to blue). Right: Example of unsupervised data (all data plotted in blue). In each case, $\mathbf{x}[i]$ is a two-dimensional vector that corresponds to a point on the plane

Typical examples of supervised data are **MNIST** [18] and **CIFAR-10** [19], which are in the format of

$$\{(\text{image data}[i], \text{label}[i])\}_{i=1,2,\ldots,\#} \tag{2.1}$$

For example, in MNIST[2] (left pictures of Fig. 2.2), data of a handwritten number is stored as 28×28 pixel image data, which can be regarded as $28 \times 28 = 784$-dimensional real vector $\mathbf{x}$. The teaching signal indicates which number $\mathbf{x}$ represents: $n \in \{0, 1, 2, 3, 4, 5, 6, 7, 8, 9\}$, or a 10-dimensional vector $\mathbf{d} = (d^0, d^1, d^2, d^3, d^4, d^5, d^6, d^7, d^8, d^9)$, where the component corresponding to the number n is $d^n = 1$ and the other components are 0.

CIFAR-10[3] (right pictures of Fig. 2.2) stores colored natural images as image data of 32×32 pixels. Each image is described by the intensity of red, green, and blue as real values for each pixel, so it can be regarded as a $32 \times 32 \times 3 = 3072$ dimensional vector $\mathbf{x}$. The teaching signal indicates whether the image $\mathbf{x}$ represents {airplane, car, bird, cat, deer, dog, frog, horse, ship, truck}, and is stored in the same expression as MNIST.

[2]Modified NIST (MNIST) is based on a database of handwritten character images created by the National Institute of Standards and Technology (NIST).

[3]A database created by the Canadian Institute for Advanced Research (CIFAR). "10" in "CIFAR-10" indicates that there are 10 teacher labels. There is also data with a more detailed label, CIFAR-100.

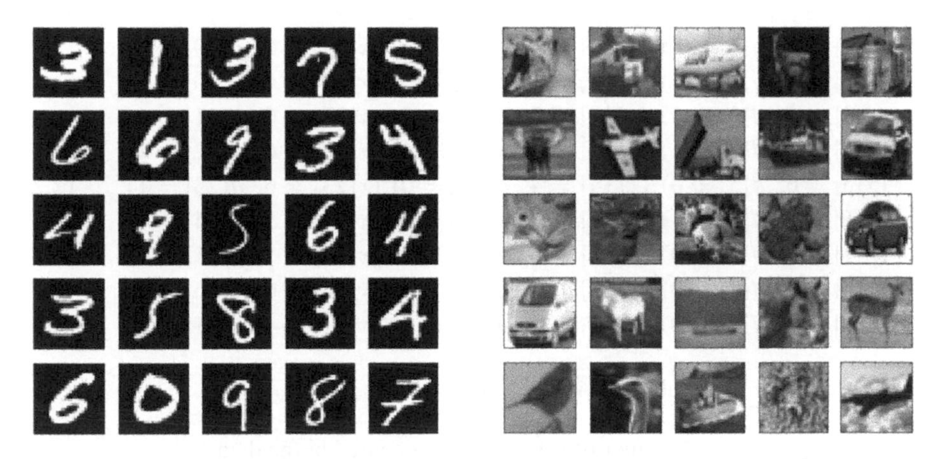

Fig. 2.2 Left: a sample in MNIST [18]. Right: a sample in CIFAR-10 [19]

2.1.1 Mathematical Formulation of Data

Given supervised/unsupervised data, the goal is to design a learning machine that extracts its features and is able to predict properties of unknown data. To find a way to construct such a machine, we need to mathematically formulate what we mean by "predicting properties of unknown data." For this reason, let us try the following thought experiment.

Two dice
Suppose we have two dice here.

- Dice A rolls every number $1, \cdots, 6$ with a probability of $1/6$.
- Dice B has only the number 6 with probability 1.

Then we repeat the following steps:

1. Choose A or B with a probability of 1/2, set $d = 0$ for A, $d = 1$ for B.
2. Roll the dice and name the pip as x.
3. Record (x, d).

Then we may obtain the following data.

$$
\begin{array}{l|l|l}
(x[1]=6,\ d[1]=1) & (x[10]=6,\ d[10]=1) & (x[19]=6,\ d[19]=1) \\
(x[2]=6,\ d[2]=1) & (x[11]=3,\ d[11]=0) & (x[20]=6,\ d[20]=1) \\
(x[3]=6,\ d[3]=1) & (x[12]=3,\ d[12]=0) & (x[21]=6,\ d[21]=1) \\
(x[4]=6,\ d[4]=1) & (x[13]=6,\ d[13]=1) & (x[22]=6,\ d[22]=1) \\
(x[5]=6,\ d[5]=0) & (x[14]=4,\ d[14]=0) & (x[23]=6,\ d[23]=1) \\
(x[6]=6,\ d[6]=1) & (x[15]=3,\ d[15]=0) & (x[24]=5,\ d[24]=0) \\
(x[7]=6,\ d[7]=0) & (x[16]=6,\ d[16]=0) & (x[25]=5,\ d[25]=0) \\
(x[8]=2,\ d[8]=0) & (x[17]=1,\ d[17]=0) & (x[26]=6,\ d[26]=1) \\
(x[9]=5,\ d[9]=0) & (x[18]=6,\ d[18]=1) & (x[27]=4,\ d[27]=0)
\end{array}
\tag{2.2}
$$

Now we define a probability with "^" (hat):

$$
\hat{P}(x,d) = \frac{\text{Number of times } (x,d) \text{ appeared in the data}}{\text{Total number of data}}\,. \tag{2.3}
$$

An example experiment taking 1000 data calculates the value of $\hat{P}(x,d)$ as follows:

$\hat{P}(x,d) =$ (2.4)

	$d=0$	$d=1$
$x=1$	0.086	0.0
$x=2$	0.087	0.0
$x=3$	0.074	0.0
$x=4$	0.089	0.0
$x=5$	0.083	0.0
$x=6$	0.082	0.499

The more data we have, the closer our table is to

$P(x,d) =$

	$d=0$	$d=1$
$x=1$	1/12	0
$x=2$	1/12	0
$x=3$	1/12	0
$x=4$	1/12	0
$x=5$	1/12	0
$x=6$	1/12	1/2

, $(1/12 = 0.08\dot{3},\ 1/2 = 0.5)$. (2.5)

In fact, the definition shows that (2.5) is the realization probability of (x,d). The data in (2.2) can be regarded as a sampling from $P(x,d)$. Let us write it as follows:

$$
(x[i], d[i]) \sim P(x,d)\,. \tag{2.6}
$$

Data generation probability known only to God
The above (2.2) is an example of the supervised data $\{(\mathbf{x}[i], \mathbf{d}[i])\}_{i=1,2,\ldots\#}$, although small in size. Supervised data such as MNIST and CIFAR-10 are basically obtained by repeating the following protocols:[4]

1. Choose a label with some probability and express it as $\mathbf{d}$.
2. Take the image corresponding to $\mathbf{d}$ by some method and set it as $\mathbf{x}$.
3. Record $(\mathbf{x}, \mathbf{d})$.

Then, it is natural to expect that behind all the data in the world there exists a data generation probability $P(\mathbf{x}, \mathbf{d})$ like (2.5), and that the data itself is the result of sampling according to the probability

$$(\mathbf{x}[i], \mathbf{d}[i]) \sim P(\mathbf{x}, \mathbf{d}) . \tag{2.7}$$

We assume the existence of the **data generation probability** $P(\mathbf{x}, \mathbf{d})$ and that (2.7) on the data is the starting point for statistical machine learning. It is said that Einstein wrotes in a letter to Born that "God does not roll the dice," but machine learning conversely says "God rolls dice." Of course, nobody knows the concrete expression of $P(\mathbf{x}, \mathbf{d})$ like (2.5), but it is useful in later discussions to make this assumption.[5]

2.2 Machine Learning and Occam's Razor

We asserted above that "we do not know the concrete expression of $P(\mathbf{x}, \mathbf{d})$." However, the goal in machine learning is to "approximately" know the concrete expression of $P(\mathbf{x}, \mathbf{d})$, although they appear to contradict each other. Basically, we

[4]This process corresponds to $P(\mathbf{x}, \mathbf{d}) = P(\mathbf{x}|\mathbf{d})P(\mathbf{d})$, but in reality the following order is easier to collect data:

1. Take an image $\mathbf{x}$.
2. Judge the label of the image and set it to $\mathbf{d}$.
3. Record $(\mathbf{x}, \mathbf{d})$.

This process corresponds to $P(\mathbf{x}, \mathbf{d}) = P(\mathbf{d}|\mathbf{x})P(\mathbf{x})$. The resulting sampling should be the same from Bayes' theorem (see the column in this chapter), admitting the existence of data generation probabilities.

[5]Needless to say, the probabilities here are all classical, and quantum theory has nothing to do with it.

take the following strategy:

I. Define the probability $Q_J(\mathbf{x}, \mathbf{d})$ depending on the parameter J

II. Adjust J to make $Q_J(\mathbf{x}, \mathbf{d})$ as close as possible to $P(\mathbf{x}, \mathbf{d})$

Stage II is called **training** or **learning**.
Suppose the model $Q_J(\mathbf{x}, \mathbf{d})$ is defined in some way. In order to bring this closer to $P(\mathbf{x}, \mathbf{d})$ at stage II, we need a function that measures the difference between them. A suitable one is **relative entropy**:[6]

$$D_{KL}(P||Q_J) = \sum_{\mathbf{x},d} P(\mathbf{x}, d) \log \frac{P(\mathbf{x}, d)}{Q_J(\mathbf{x}, d)} . \tag{2.8}$$

Many readers may be unfamiliar with the $||$ notation used here. This is just a notation to emphasize that the arguments P, Q_J are asymmetric. The relative entropy is non-negative, i.e., $D_{KL} \geq 0$, where the equality holds if and only if $P(\mathbf{x}, d) = Q_J(\mathbf{x}, d)$ for any $(\mathbf{x}, d)$ (see the column of this chapter). Therefore, if we can adjust (learn) J and reduce D_{KL}, learning will proceed. Equation (2.8) is called **generalization error**.[7] **Generalization** is a technical term used to describe the state in which a machine can adapt to unknown situations, as described at the beginning of this chapter.

Importance of the number of data
Actual learning has various difficulties. First of all, it is impossible to calculate the value or gradient of (2.8) because we never know the specific expression of $P(\mathbf{x}, \mathbf{d})$. In realistic situations, we use the approximate probability $\hat{P}(\mathbf{x}, \mathbf{d})$ derived from the data (this is called **empirical probability**) such as (2.4), and consider

$$D_{KL}(\hat{P}||Q_J) \tag{2.9}$$

which is called **empirical error**.[8] From this, it can be intuitively understood that reliable learning results cannot be obtained unless the number of data is large enough. For example, (2.4), the calculation was performed with a set of 1000 data.

[6] As mentioned in Chap. 1, relative entropy is also called **Kullback–Leibler divergence**. Although it measures the "distance," it does not satisfy the axiom of symmetry for the distance, so it is called **divergence**.

[7] In general, "generalization error" often refers to the expectation value of the error function (which we will describe later). As shown later, they are essentially the same thing.

[8] This is the same as using maximum likelihood estimation, just as we used it when we introduced relative entropy in Chap. 1.

If we use just 10 data instead, we obtain

$$\hat{P}(x,d) = \begin{array}{|c|c|c|}\hline & d=0 & d=1 \\ \hline x=1 & 0.0 & 0.0 \\ \hline x=2 & 0.0 & 0.0 \\ \hline x=3 & 0.1 & 0.0 \\ \hline x=4 & 0.1 & 0.0 \\ \hline x=5 & 0.1 & 0.0 \\ \hline x=6 & 0.2 & 0.5 \\ \hline \end{array} \tag{2.10}$$

One finds that values that should be non-zero are zero, and that it is far from the true probability (2.5). In this case, the model $Q_J(x,d)$ can be modeled only on (2.10), so the accuracy for an approximation of the true probability (2.2) is worse.

2.2.1 *Generalization*

We have to keep in mind that the ultimate goal was to make (2.8) as close to zero as possible, even though what we really can calculate is only (2.9). From that point of view, solving the problem of making (2.9) vanish may not always be helpful. To make matters worse, reducing the empirical error (2.9) to an extremely small value often causes a phenomenon called **over-training**, in which the value of the generalization error (2.8) increases even though the value of the empirical error is small. This is something similar to the following daily experience: even if one gets a perfect score in a regular test by memorizing every word in one's textbook, one has not reached real understanding, so one gets a low score in an achievement test.[9] Rather, it is better to consider reducing (2.8) only from the information of (2.9). If this is achieved, Q_J will be sufficient as an approximation of the data generation probability P, and sampling from Q_J will have the generalization ability, the ability to "predict about unknown data."

Difficulty of generalization

In fact, by using the above definitions, one can show the inequality [21]:[10]

$$(\text{Generalization error}) \leq (\text{Experience error}) + O\Big(\sqrt{\frac{\log(\#/d_{VC})}{\#/d_{VC}}}\Big). \tag{2.11}$$

[9] If the reader knows experimental physics, recall overfitting. See also [20].

[10] As described in the footnote below, this holds when the story is limited to binary classification. In addition, it is an inequality that can be used when both the generalization error and the empirical error have been normalized to take a value of [0, 1] by some method [21].

Here d_{VC} is a numerical value of the expressive power (or complexity) of the model Q_J called the **VC dimension** (Vapnik-Chervonenkis dimension). Roughly speaking, d_{VC} corresponds to the number of parameters J included in the model.[11] Let us consider the difficulty of generalization using this expression.

First, adjust the parameter J to reduce the empirical error, since it is the first thing that can be done. It is better to increase the expressive power of the model, because otherwise it cannot handle various data. As a result, in principle, empirical errors could be very small by complicating the model.

On the other hand, the value of d_{VC} jumps up in a complex model that makes the empirical error extremely small. For simplicity, let us say this has become infinite. Then, under the fixed number of data #, the value of the second term in (2.11) diverges as

$$\lim_{\#/d_{VC} \to +0} \frac{\log(\#/d_{VC})}{\#/d_{VC}} \to \infty \,. \tag{2.14}$$

Then (2.11) is a meaningless inequality

$$\text{(Generalization error)} \leq \text{(Experience error)} + \infty. \tag{2.15}$$

No matter how much the empirical error is reduced in this state, it is not guaranteed that the generalization error is reduced.[12] Therefore, generalization of learning cannot be guaranteed without using an appropriate model according to the scale of the data. When creating a phenomenological model of physics, models that are too complicated to describe phenomena are not preferable. This argument is called **Occam's razor**. The inequality (2.11) mathematically expresses the principle of Occam's razor.[13]

---

[11] The exact definition of the VC dimension is as follows. We define the model Q_J as

$$Q_J(\mathbf{x}, \mathbf{d}) = Q_J(\mathbf{d}|\mathbf{x})P(\mathbf{x}) \,, \tag{2.12}$$

as we will do later. Also, using function f_J with parameter J and the Dirac delta function δ, we write

$$Q_J(\mathbf{d}|\mathbf{x}) = \delta(f_J(\mathbf{x}) - \mathbf{d}). \tag{2.13}$$

Suppose further that we are working on the problem of binary classification with $\mathbf{d} = 0, 1$. It means that $f_J(\mathbf{x})$ is working to assign the input $\mathbf{x}$ to either 0 or 1. By the way, if there are # data, there are $2^{\#}$ possible ways of assigning 0/1. If we can vary J fully, then $[f_J(\mathbf{x}[1]), f_J(\mathbf{x}[2]), \ldots, f_j(\mathbf{x}[\#])]$ can realize all possible 0/1 distributions, and this model has the ability to completely fit the data (the capacity is saturated compared to the number of data #). The VC dimension refers to the maximum value of # where such a situation is realized.

[12] Over-training is a situation that falls into this state.

[13] A well-known index based on a similar idea is **Akaike's Information Criteria** (AIC) [22]:

$$AIC = -2\log L + 2k \,, \tag{2.16}$$

The mystery of generalization in deep learning

On the other hand, as explained below, models in deep learning become more and more complex with the number of layers. For example, the famous ResNet [24] has hundreds of layers, which gives an incredibly huge d_{VC}.[14] Therefore, from the inequality of (2.11), it cannot be guaranteed that the generalization performance is improved by the above logic. Nevertheless, ResNet appears to have gained strong generalization performance. This appears to contradict with the above logic, but it does not, because although the evaluation using (2.11) cannot guarantee generalization performance, more detailed inequalities may exist:

$$\text{(Generalization error)} \underbrace{\overset{?}{\leq} \text{(Experience error)} + O_{DL}}_{\text{something like this inside (2.11)?}}$$

$$\leq \text{(Experience error)} + O\Big(\sqrt{\frac{\log(\#/d_{VC})}{\#/d_{VC}}}\Big). \tag{2.17}$$

There is a movement to solve the mystery of the generalization performance of deep learning from such more detailed inequality evaluation (for example, [27] etc.), but there is no definitive result at the time of writing (as of January 2019).

2.3 Stochastic Gradient Descent Method

One of the easiest ways to implement learning is to use the gradient of the error parameter J (the second term of the following equation) and update the parameters as

$$J_{t+1} = J_t - \epsilon \nabla_J D_{KL}(P||Q_J), \tag{2.18}$$

Where L is the maximum likelihood and k is the number of model parameters. Also, the AIC and the amount of χ^2 that determine the accuracy of fitting have a relationship [23].

[14] For example, according to the theorem 20.6 of [25], if the number of learning parameters of a neural network having a simple step function as an activation function (described in the next section) is N_J, the VC dimension of the neural network is the order of $N_J \log N_J$. ResNet is not such a neural network, but let us estimate its VC dimension with this formula for reference. For example, according to Table 6 of [24], a ResNet with 110 layers (=1700 parameters) has an average error rate of 6.61% for the classification of CIFAR-10 (60,000 data), while the VC dimension is 12,645.25 according to the above formula, and the second term of (2.11) is 0.57. Since the errors are scaled to [0,1] in the inequalities, the error rate can be read as at most about 10%, and the above-mentioned error rate of 6.61% overwhelms this. A classification error of ImageNet [26] (with approximately 10^7 data) is written in table 4 of the same paper, and this has a top-5 error rate of 4.49% in 152 layers, while the same simple calculation gives about 9%, so the reality is still better than the upper limit of the inequality. The inequality (2.11) is a formula for binary classification, but CIFAR-10 has 10 classes and ImageNet has 1000 classes, so we should consider the estimation here as a reference only.

after setting an appropriate initial value $J_{t=0}$. This is called the **gradient descent method**.[15] Here ϵ is a small positive real number. Then, for a change of J as $\delta J = J_{t+1} - J_t$, the change of D_{KL} is

$$\delta D_{KL}(P||Q_J) \approx \delta J \cdot \nabla_J D_{KL}(P||Q_J) = -\epsilon |\nabla_J D_{KL}(P||Q_J)|^2, \tag{2.19}$$

and we can see that D_{KL} decreases.[16] However, as we have emphasized many times, calculating the gradient requires knowing the true data distribution P, so this is an armchair theory. The problem is, how we can get closer to (2.18)?

Stochastic gradient descent method

So how can we make quantities that give a good approximation of (2.18)? First, let us calculate the slope of the error,

$$\begin{aligned} \nabla_J D_{KL}(P||Q_J) &= \nabla_J \sum_{\mathbf{x},d} P(\mathbf{x},d) \log \frac{P(\mathbf{x},d)}{Q_J(\mathbf{x},d)} \\ &= \sum_{\mathbf{x},d} P(\mathbf{x},d) \Big(\nabla_J \log \frac{P(\mathbf{x},d)}{Q_J(\mathbf{x},d)} \Big). \end{aligned} \tag{2.20}$$

Thus, the gradient of the error is the expectation value of $\nabla_J \log \frac{P(\mathbf{x},d)}{Q_J(\mathbf{x},d)}$. So, we consider approximating the expectation value with the sample average from P,

$$(\mathbf{x}[i], d[i]) \sim P(\mathbf{x},d), \quad i = 1, 2, \ldots, \#, \tag{2.21}$$

$$\hat{g} = \sum_{i=1}^{\#} \frac{1}{\#} \Big(\nabla_J \log \frac{P(\mathbf{x}[i], d[i])}{Q_J(\mathbf{x}[i], d[i])} \Big). \tag{2.22}$$

This quantity will certainly reduce to (2.20) once the expectation value is taken for each sample. Also, due to the law of large numbers, a large # should be a good approximation of (2.20). We call the gradient descent method using the sample approximation the **stochastic gradient descent** (SGD) method [28]:

1. Initialize $J_{t=0}$ properly
2. Repeat the following (repetition variable is t):

   Sample # data (see (2.21))

[15] In fact, the first derivative $\nabla_J D_{KL}(P||Q_J)$ is not enough to minimize the error. A Hessian corresponding to the second derivative is what we should look at, but it is not practical because of the computational complexity.

[16] If the value of ϵ is too large, the approximation "$\approx$" in the expression (2.19) will be poor, and the actual parameter update will behave unintentionally. This is related to the gradient explosion problem described in Chap. 4.

Calculate $\hat{g}$ according to the formula (2.22)

$$J_{t+1} = J_t - \epsilon \hat{g}. \tag{2.23}$$

Note that in order to justify this approximation using the law of large numbers, all sampling in (2.21) must be taken independently, so it must be performed separately at each step t of the parameter update. Such a learning method that always uses new data at each t is called online learning.

When data is limited

When a prepared database is used, the above is not possible. In that case, a method called batch learning is used. We want to consider a stochastic gradient descent method that also minimizes the bias, and in deep learning, the following method called mini-batch learning is often used:

(Data$\{(\mathbf{x}[i], d[i])\}_{i=1,2,\dots\#}$ shall be given in advance)

1. Initialize J properly.
2. Repeat the following:

 Divide data into$M$ partial data randomly.

 Repeat the following from $m = 1$ to $m = M$:

 Replace data in (2.22) with the mth partial data and calculate $\hat{g}$.

$$J \leftarrow J - \epsilon \hat{g}. \tag{2.24}$$

Note that at the beginning of the loop in step 2, there is a process for randomly splitting the data. This method is mostly used when applying the gradient descent method in supervised learning. Unlike the original gradient descent method (2.23), even if the loop of step 2 is repeated, each $\hat{g}$ is not independent of the others, thus there is no guarantee that the approximation accuracy of the gradient of the generalization error will be better. Normally, it is necessary to prepare **validation data** separately and monitor (observe) the empirical error at the same time, so as to avoid over-training.

The case of deep learning

In most cases, the training is done according to the method (2.24). Here are some points to emphasize. First, the gradient $\hat{g}$ is the gradient of the **error function** which will be introduced later, and when a neural network is used, the differential calculation can be algorithmized: it is called **back-propagation**, and provides a fast algorithm which will be explained later. Also, be aware that not only the simple stochastic gradient descent method but also various evolved forms of the gradient descent method are often used. References that summarize various gradient methods include [28]. In addition, original papers (listed in [28]) can be read for free.

Column: Probability Theory and Information Theory

As explained in this book, probability theory and information theory are indispensable for developing the theory of machine learning. This column explains some important concepts.

Joint and Conditional Probabilities

Machine learning methods often consider the probability of the input value $\mathbf{x}$ and the teaching signal $\mathbf{d}$. The **joint probability** represents the appearance probability of these two variables. Let us write it as

$$P(\mathbf{x}, \mathbf{d})\,. \tag{2.25}$$

Of course, we have

$$1 = \sum_{\mathbf{x}} \sum_{\mathbf{d}} P(\mathbf{x}, \mathbf{d})\,. \tag{2.26}$$

Here, $\sum$ is a sum when $\mathbf{x}$, $\mathbf{d}$ are discrete valued, and is replaced by an integral when they are continuous variables.

There should be some situations when we do not want to look at the value of $\mathbf{x}$ but want to consider the probability of $\mathbf{d}$. It can be expressed as

$$P(\mathbf{d}) = \sum_{\mathbf{x}} P(\mathbf{x}, \mathbf{d}). \tag{2.27}$$

The summation operation is called marginalizing, and the probability $P(\mathbf{d})$ is called **marginal probability**.

In some cases, $\mathbf{d}$ may already be given. This situation is represented by

$$P(\mathbf{x}|\mathbf{d})\text{: probability of } \mathbf{x} \text{ when } \mathbf{d} \text{ is given.} \tag{2.28}$$

This is called **conditional probability**. In this case, $\mathbf{d}$ is regarded as a fixed quantity,

$$1 = \sum_{\mathbf{x}} P(\mathbf{x}|\mathbf{d})\,. \tag{2.29}$$

So, in fact, we find

$$P(\mathbf{x}|\mathbf{d}) = \frac{P(\mathbf{x}, \mathbf{d})}{P(\mathbf{d})} = \frac{P(\mathbf{x}, \mathbf{d})}{\sum_{\mathbf{X}} P(\mathbf{X}, \mathbf{d})}\,. \tag{2.30}$$

The numerator uses the original joint probability, but this equality can be considered as a renormalization of it due to the fixing of **d**. If you are familiar with statistical mechanics of spins under an external magnetic field, this concept can be understood by considering the spin degrees of freedom as **x** and the external magnetic field as **d**. By the way, even if the roles of **x** and **d** are exchanged, the same holds for the same reason:

$$P(\mathbf{d}|\mathbf{x}) = \frac{P(\mathbf{x}, \mathbf{d})}{P(\mathbf{x})} = \frac{P(\mathbf{x}, \mathbf{d})}{\sum_{\mathbf{D}} P(\mathbf{x}, \mathbf{D})}. \tag{2.31}$$

Note that the numerator on the right-hand side is common between (2.30) and (2.31),

$$P(\mathbf{x}|\mathbf{d})P(\mathbf{d}) = P(\mathbf{d}|\mathbf{x})P(\mathbf{x}) = P(\mathbf{x}, \mathbf{d}). \tag{2.32}$$

This is called **Bayes' theorem** and is often used in this book.

Monty Hall problem and conditional probability

Here is a notorious problem of probability theory. Suppose only one of three boxes has a prize. Let the contents of each box be x_1, x_2, x_3. Possible combinations are

$$\begin{aligned}
&(x_1 = \circ,\ x_2 = \times,\ x_3 = \times), \\
&(x_1 = \times,\ x_2 = \circ,\ x_3 = \times), \\
&(x_1 = \times,\ x_2 = \times,\ x_3 = \circ).
\end{aligned} \tag{2.33}$$

So it is a good idea to take the probability of $1/3$ for each.

$$P(x_1, x_2, x_3) = \frac{1}{3}. \tag{2.34}$$

Then you pick any box. At this stage, all boxes are equal, so suppose you pick x_1. The probability of finding the prize is $1/3$, the probability of not finding it is $2/3$:

$$P(x_1 = \circ) = \frac{1}{3}, \quad P(x_1 = \times) = \frac{2}{3}. \tag{2.35}$$

This is just a marginal probability because you have summed the probability other than x_1.

Now, suppose you can't open the x_1 box. Instead, there is a moderator, who knows the contents of every box. The moderator opens one of the boxes x_2, x_3 and shows to you that there is nothing inside. Then, all that remain are x_1, which you chose, and one of x_2, x_3, which the moderator did not open. Should you stay at x_1 or choose the box that the moderator did not open? This is the Monty Hall problem. Let us solve this problem taking into account conditional probabilities. First,

$$P(x_2 \text{ or } x_3 = |x_1 = \circ) = 0. \tag{2.36}$$

It is evident in this setting that only one box has the prize. On the other hand,

$$
\begin{aligned}
&P(x_2 \text{ or } x_3 = |x_1 = \times) \\
&= \underbrace{P\Big((x_2, x_3) = (\circ, \times)\Big|x_1 = \times\Big)}_{\frac{P\left((x_1, x_2, x_3) = (\times, \circ, \times)\right)}{P(x_1 = \times)}} + \underbrace{P\Big((x_2, x_3) = (\times, \circ)\Big|x_1 = \times\Big)}_{\frac{P\left((x_1, x_2, x_3) = (\times, \times, \circ)\right)}{P(x_1 = \times)}} \\
&= \frac{1/3}{2/3} + \frac{1/3}{2/3} = \frac{1}{2} + \frac{1}{2} = 1\,.
\end{aligned} \tag{2.37}
$$

And in this case, it is guaranteed that you get the prize, which is consistent with our intuition. Then, the probability of winning the prize by changing the box selection is

$$
\begin{aligned}
&P(\text{With the change} = \circ) \\
&= \underbrace{P(x_2 \text{ or } x_3 = \circ|x_1 = \circ)}_{0}\,\underbrace{P(x_1 = \circ)}_{1/3} + \underbrace{P(x_2 \text{ or } x_3 = \circ|x_1 = \times)}_{1}\,\underbrace{P(x_1 = \times)}_{2/3} \\
&= \frac{2}{3}.
\end{aligned} \tag{2.38}
$$

The probability of finding the prize is higher than $P(x_1 = \circ) = 1/3$ when not changing, so the answer to this problem is "you should change." More simply, combining

$$
P(\text{Without the change} = \circ) = P(x_1 = \circ) = \frac{1}{3}\,, \tag{2.39}
$$

and

$$
P(\text{With the change} = \circ) + P(\text{Without the change} = \circ) = 1\,, \tag{2.40}
$$

you find

$$
P(\text{With the change} = \circ) = \frac{2}{3}\,. \tag{2.41}
$$

This is a simple problem, but at first glance, the probability of winning is $1/3$ and it does not change whatever the moderator does, so it seems that the result is the same whether or not you change the box. Certainly, the moderator does not interfere with the original probability $1/3$, but when information is added, we should consider the conditional probability instead of the original probability.[17] This is a good problem in the sense that it teaches us.

Relative Entropy

We introduced relative entropy D_{KL} in this chapter. This quantity is one of the most important functions in the field of machine learning. Here, we will show the properties quoted in this chapter: for a probability distribution $p(x), q(x)$, we have $D_{KL}(p||q) \geq 0$, and only when $D_{KL}(p||q) = 0$ the probability distributions match, $p(x) = q(x)$. In addition, as an exercise, we will explain a calculation example using the Gaussian distribution and some unexpected geometric structure behind it.

Definition and properties

For probability distributions $p(x)$ and $q(x)$, we define

$$D_{KL}(p||q) := \int dx\ p(x) \log \frac{p(x)}{q(x)} \tag{2.42}$$

and call it **relative entropy**. The relative entropy satisfies

$$D_{KL}(p||q) \geq 0, \quad D_{KL}(p||q) = 0 \Leftrightarrow \forall x,\ p(x) = q(x). \tag{2.43}$$

There are several ways to check the above properties, and in this column we prove them in the following way. First, write the relative entropy as

$$D_{KL}(p||q) = \int dx \left[p(x) \log \frac{p(x)}{q(x)} + q(x) - p(x) \right]. \tag{2.44}$$

Here we used $\int dx\ q(x) = \int dx\ p(x) = 1$. We simply added unity and subtracted unity. And we can transform it as

$$(2.44) = \int dx\ p(x) \left[\left(\frac{q(x)}{p(x)} - 1 \right) - \log \frac{q(x)}{p(x)} \right]. \tag{2.45}$$

Using (see Fig. 2.3)

$$(X - 1) \geq \log X\,, \tag{2.46}$$

[17] For example, even if you have a favorite restaurant in your neighborhood, if you hear rumors that the food at another restaurant is really delicious, then you will be tempted to go there. Considering the conditional probability is something close to this simple feeling.

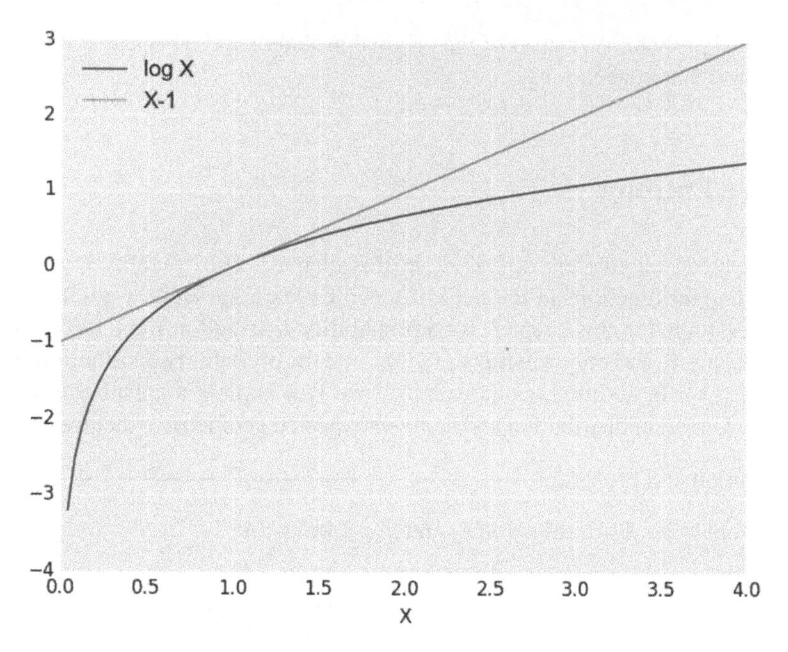

Fig. 2.3 Green: $(X - 1)$, Blue: graph of $\log X$

the integrand of (2.45) is always equal to or larger than zero. Since the integrand is always equal to or larger than zero, we can show

$$D_{KL}(p||q) \geq 0. \tag{2.47}$$

Next, the equality is satisfied when the equality of (2.46) is established ($X = 1$). This means that for any x we need to have

$$\frac{q(x)}{p(x)} = 1, \tag{2.48}$$

which shows (2.43).

Example: Gaussian distribution
To get a sense of the relative entropy, let us calculate the relative entropy for the **Gaussian distribution**:

$$\frac{1}{\sqrt{2\pi}\sigma}e^{-\frac{1}{2\sigma^2}(x-\mu)^2}. \tag{2.49}$$

So here we consider

$$p(x) = \frac{1}{\sqrt{2\pi}\sigma_p} e^{-\frac{1}{2\sigma_p^2}(x-\mu_p)^2} , \tag{2.50}$$

$$q(x) = \frac{1}{\sqrt{2\pi}\sigma_q} e^{-\frac{1}{2\sigma_q^2}(x-\mu_q)^2} . \tag{2.51}$$

Then by definition

$$\begin{aligned}
D_{KL}(p||q) &= \int_{-\infty}^{\infty} dx\ p(x)\Big(\log \frac{\sigma_q}{\sigma_p} - \frac{1}{2\sigma_p^2}(x-\mu_p)^2 + \frac{1}{2\sigma_q^2}(x-\mu_q)^2 \Big) \\
&= \log \frac{\sigma_q}{\sigma_p} - \frac{1}{2\sigma_p^2}(\sigma_p)^2 + \frac{1}{2\sigma_q^2}\int_{-\infty}^{\infty} dx\ p(x) \underbrace{(x-\mu_q)^2}_{(\mu_p-\mu_q)^2+2(\mu_p-\mu_q)(x-\mu_p)+(x-\mu_p)^2} \\
&= \log \frac{\sigma_q}{\sigma_p} - \frac{1}{2\sigma_p^2}(\sigma_p)^2 + \frac{1}{2\sigma_q^2}\Big((\mu_p-\mu_q)^2 + 0 + \sigma_p^2 \Big) \\
&= \frac{1}{2}\left(-\log \frac{\sigma_p^2}{\sigma_q^2} + \left(\frac{\sigma_p^2}{\sigma_q^2} - 1 \right) + \frac{1}{\sigma_q^2}(\mu_p-\mu_q)^2 \right)
\end{aligned} \tag{2.52}$$

At the second equality we made a Taylor expansion. You can again show that this value is positive, using (2.46).

Gaussian distribution and AdS spacetime

The definition of relative entropy seems to represent the "distance" between probability distributions. If there is a distance, each probability distribution corresponds to a point, and when it is collected, it becomes a "space."[18] In fact, in the Gaussian distribution example above, the mean μ takes a value between $(-\infty, +\infty)$ and σ takes a value between $(0, \infty)$ so the whole Gaussian distributions can form a "space" $(-\infty, +\infty) \times (0, +\infty)$. Let us find the "infinitesimal distance" in this case from the relative entropy. We consider a point

$$\sigma_p = \sigma , \quad \mu_p = \mu , \tag{2.53}$$

and a point close to it

$$\sigma_q = \sigma + d\sigma , \quad \mu_q = \mu + d\mu . \tag{2.54}$$

[18] In the research field called information geometry, this viewpoint is used.

We calculate the relative entropy up to the second order in $d\sigma, d\mu$ as

$$\begin{aligned}D_{KL}(p||q) &= \frac{1}{2}\left(-\log\frac{\sigma^2}{(\sigma+d\sigma)^2}+\left(\frac{\sigma^2}{(\sigma+d\sigma)^2}-1\right)+\frac{1}{(\sigma+d\sigma)^2}(\mu-\mu-d\mu)^2\right)\\ &= \frac{1}{2}\left(\log(1+\frac{d\sigma}{\sigma})^2+\left(\frac{1}{(1+\frac{d\sigma}{\sigma})^2}-1\right)+\frac{1}{\sigma^2(1+\frac{d\sigma}{\sigma})^2}d\mu^2\right)\\ &\approx \frac{1}{2}\left(2\frac{d\sigma}{\sigma}-\frac{d\sigma^2}{\sigma^2}+\left(1-2\frac{d\sigma}{\sigma}+3\frac{d\sigma^2}{\sigma^2}-1\right)+\frac{d\mu^2}{\sigma^2}\right)\\ &= \frac{1}{2}\left(2\frac{d\sigma^2}{\sigma^2}+\frac{d\mu^2}{\sigma^2}\right)=\frac{d\sigma^2+d\tilde{\mu}^2}{\sigma^2}. \end{aligned} \tag{2.55}$$

(In the last expression, we defined $\tilde{\mu}=\mu/\sqrt{2}$.) This is the metric of a hyperboloid, and a part of the metric of the **anti-de Sitter spacetime** (AdS spacetime). In recent years, the anti-de Sitter spacetime has played an important role in phenomena called **AdS/CFT correspondence**[19] in string theory. It is interesting that Gaussian space also has such a "hidden" anti-de Sitter spacetime. Anti-de-Sitter spacetime has the property that the metric diverges at its infinite boundary (the distance is stretched infinitely). In the current context it corresponds to $\sigma\approx 0$. σ is the variance of the Gaussian distribution, so the origin of this anomalous behavior is clear: the limit is no longer a function in the usual sense, since the limit of bringing the variance of the Gaussian distribution to zero corresponds to the Dirac delta function. Chapter 12 introduces research relating AdS/CFT correspondence to machine learning.

[19] CFT (conformal field theory) is a special class of quantum field theory.

Chapter 3
Basics of Neural Networks

Abstract In this chapter, we derive neural networks from the viewpoint of physical models. A neural network is a nonlinear function that maps an input to an output, and giving the network is equivalent to giving a function called an error function in the case of supervised learning. By considering the output as dynamical degrees of freedom and the input as an external field, various neural networks and their deepened versions are born from simple Hamiltonians. Training (learning) is a procedure for reducing the value of the error function, and we will learn the specific method of backpropagation using the bra-ket notation popular in quantum mechanics. And we will look at how the "universal approximation theorem" works, which is why neural networks can express connections between various types of data.

Now, let us move on to the explanation of supervised learning using neural networks. Unlike many deep learning textbooks, this chapter describes machine learning in terms of classical statistical physics.[1]

3.1 Error Function from Statistical Mechanics

Error function is the log term of the error described by Kullback Leibler divergence which we introduced in (2.8) and (2.9) in the previous chapter. The general direction of explanation in many textbooks is to discuss the generalization error and empirical error after defining the error function, but in this book, we will explain that starting from (2.8) and (2.9), by considering the problem settings of **supervised learning** as an appropriate statistical mechanical system, an appropriate error function can be derived for each problem. In the process, we will see that the structure of nonlinear functions such as sigmoid functions used in deep learning

[1]For conventional explanations, we recommend Ref. [20]. Reading them together with this book may complement understanding.

A. Tanaka et al., *Deep Learning and Physics*, Mathematical Physics Studies,
https://doi.org/10.1007/978-981-33-6108-9_3

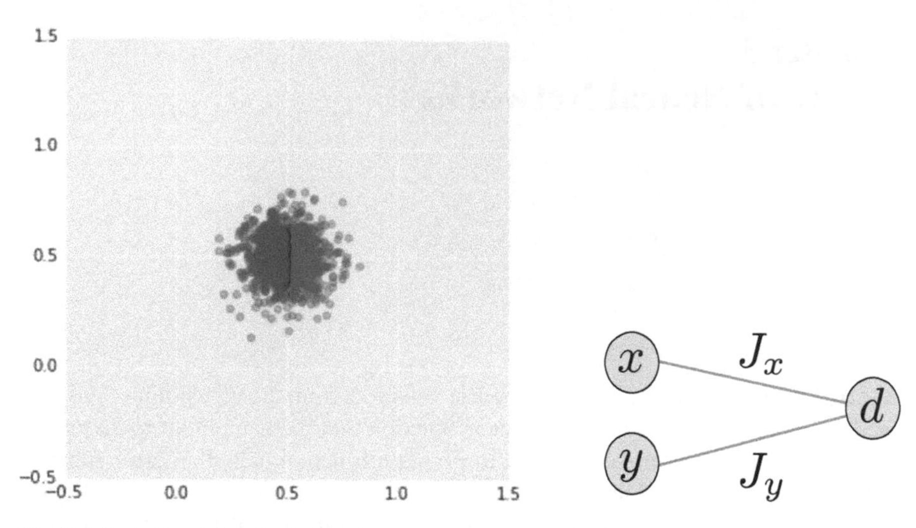

Fig. 3.1 Left: Example of supervised data ($d[i] = 0$ corresponds to red, $d[i] = 1$ corresponds to blue). Right: Diagram of the model Hamiltonian (3.6)

appear naturally. Furthermore, the origin of the rectified linear unit (ReLU) function often used in deep learning will be explained from the standpoint of statistical mechanics, providing a foothold for deepening.

3.1.1 From Hamiltonian to Neural Network

Binary classification
Let us look at the example of the supervised data given earlier. In Fig. 3.1, it appears that $d = 0$ or $d = 1$ depends on whether the value of x of $\mathbf{x} = (x, y)$ exceeds 0.5 or not. In other words, $\mathbf{x}$ and d seem to be correlated.

We shall consider expressing such a correlation with a physical system. Any physical system is defined by a **Hamiltonian**, and the Hamiltonian is written as a function of dynamical degrees of freedom.[2] When there are several dynamical degrees of freedom $A_i(t)(i = 1, 2, \cdots)$ in the physical system, Hamiltonian H is given by a function of those. (Precisely, $A_i(t)$ is a function, so H is a function of a function, that is, a functional). For example, if $A_1(t)$ represents the spatial coordinate $x(t)$ of a particle and $A_2(t)$ represents its momentum $p(t)$, the Hamiltonian of the harmonic oscillator is

$$H = \frac{1}{2}p(t)^2 + \frac{1}{2}x(t)^2 . \tag{3.1}$$

[2]If the readers are new to analytical mechanics, there is no problem in translating mechanical degrees of freedom into coordinates (or momentum) and Hamiltonians into energy.

Here, the mass of the particle and the frequency are set as unity for simplicity. If there are two particles, it is natural to have the Hamiltonian expressed as the sum of each,

$$H = \frac{1}{2}p_1(t)^2 + \frac{1}{2}x_1(t)^2 + \frac{1}{2}p_2(t)^2 + \frac{1}{2}x_2(t)^2 . \tag{3.2}$$

However, in this case, the two particles are not correlated. Each behaves as a free particle; the energy, or Hamiltonian, is simply the sum of their energies. Two particles are correlated, for example, when a force acts between the first particle and the second particle. If the magnitude of this force is proportional to the distance between the two particles, the Hamiltonian has an additional term,

$$\Delta H = c\,(x_1(t) - x_2(t))^2 . \tag{3.3}$$

Expanding the parentheses, we find it contains $cx_1(t)x_2(t)$ expressed as the product of the dynamical degrees of freedom of the two particles. This represents the correlation between the two. To summarize, correlation is generally indicated by a term composed of multiplication of dynamical degrees of freedom in the Hamiltonian. The strength of the correlation is represented by the product term coefficient c. This coefficient is called a **coupling constant**.

Now, given a physical system, how can its statistical-mechanical behavior be written in general? When a physical system is in contact with a thermal bath with temperature T, the physical system receives energy from the thermal bath, and its energy distribution takes the form of the **Boltzmann distribution** (canonical distribution). The probability P of realizing a state with energy E is given by

$$P = \frac{1}{Z}\exp\left[-\frac{E}{k_{\mathrm{B}}T}\right] . \tag{3.4}$$

Here k_{B} is the Boltzmann constant and Z is the partition function,

$$Z = \sum_{\text{all states}} \exp\left[-\frac{H}{k_{\mathrm{B}}T}\right] . \tag{3.5}$$

This is to normalize the probability P so that the total probability is 1. For simple examples, see the end-of-chapter columns in this chapter and in Chap. 5.

After the introduction of the statistical mechanics description of physical systems, let us return to Fig. 3.1, which is an example of supervised data. Since $\mathbf{x}$ and d seem to be correlated, first, as a physical system, we consider statistical mechanics with $\mathbf{x}$ as the external field and d as the dynamical degree of freedom. For the sake of simplicity, we decided to include only the first-order term in d, and write the simplest Hamiltonian as

$$H_{J,\mathbf{x}}(d) = -(xJ_x + yJ_y + J)\,d . \tag{3.6}$$

Here J_x, J_y, and J are coupling constants. Let us make a model $Q_J(\mathbf{x}, d)$ using this "physical system." First, from (3.6), it is natural to make a Boltzmann distribution about d,[3]

$$Q_J(d|\mathbf{x}) = \frac{e^{-H_{J,\mathbf{x}}(d)}}{Z} = \frac{e^{(xJ_x+yJ_y+J)d}}{\sum_{\tilde{d}=0}^{1} e^{(x^\mu J_\mu+J)\tilde{d}}} = \frac{e^{(xJ_x+yJ_y+J)d}}{1+e^{(x^\mu J_\mu+J)}}. \tag{3.7}$$

Here, we introduce a function called the **sigmoid function**,[4]

$$\sigma(X) = \frac{1}{e^{-X}+1}. \tag{3.8}$$

Then we find

$$Q_J(d=1|\mathbf{x}) = \sigma(xJ_x + yJ_y + J), \tag{3.9}$$

$$Q_J(d=0|\mathbf{x}) = 1 - \sigma(xJ_x + yJ_y + J). \tag{3.10}$$

This expression is for a given $\mathbf{x}$, but if we suppose the existence of data generation probability $P(\mathbf{x}, d)$, the $\mathbf{x}$ should be generated according to

$$P(\mathbf{x}) = P(\mathbf{x}, d=0) + P(\mathbf{x}, d=1), \tag{3.11}$$

so we use this to define

$$Q_J(\mathbf{x}, d) = Q_J(d|\mathbf{x})P(\mathbf{x}). \tag{3.12}$$

Now that we have a model, let us consider actually adjusting (learning) the coupling constants J_x, J_y, and J to mimic the true distribution $P(\mathbf{x}, d)$. The relative entropy (2.8) is calculated from settings such as (3.12),

$$\begin{aligned} D_{KL}(P||Q_J) &= \sum_{\mathbf{x},d} P(\mathbf{x}, d) \log \frac{P(d|\mathbf{x})}{Q_J(d|\mathbf{x})} \\ &= -\sum_{\mathbf{x},d} P(\mathbf{x}, d) \log Q_J(d|\mathbf{x}) + (J\text{-independent part}). \end{aligned} \tag{3.13}$$

[3] In the Boltzmann distribution, the factor of $H/(k_B T)$, which is the Hamiltonian divided by the temperature, is in the exponent. We redefine J to $Jk_B T$ to absorb the temperature part, so that the temperature does not appear in the exponent.

[4] The reason that the sigmoid function is similar to the Fermi distribution function can be understood from the fact that d takes on a binary value in the current system. $d = 0$ corresponds to the state where the fermion site is not occupied (vacancy state), and $d = 1$ corresponds to the state where fermion excitation exists.

When we use the gradient method, we only need to pay attention to the first term,[5] and make it as small as possible. Finally, if this is approximated by data (empirical probability $\hat{P}$), using (3.9), (3.10), etc., we find:

$$\begin{aligned}
&\text{First term of (3.13)} \\
&\approx -\sum_{i:\text{Data}} \frac{1}{\#} \log Q_J(d[i] \mid \mathbf{x}[i]) \\
&= \frac{1}{\#} \sum_{i:\text{Data}} \Big(- \Big[d[i] \log \Big\{ \sigma(x[i]J_x + y[i]J_y + J) \Big\} \\
&\qquad\qquad + (1 - d[i]) \log \Big\{ 1 - \sigma(x[i]J_x + y[i]J_y + J) \Big\} \Big] \Big) . \qquad (3.14)
\end{aligned}$$

Here # means the number of data. After all, reducing this value by adjusting J is called "learning."

By the way, what is the expectation value of the output when J_x, J_y, J and input data $\mathbf{x}[i]$ are fixed? A calculation gives

$$\langle d \rangle_{J,\mathbf{x}[i]} = \sum_{d=0,1} d \cdot Q_J(d|\mathbf{x}[i]) = Q_J(d = 1|\mathbf{x}[i]) = \sigma(x[i]J_x + y[i]J_y + J) . \qquad (3.15)$$

This is considered the "output value" of the machine. Then, using the following function called **cross entropy**,

$$L(X, d) = -\Big[d \log X + (1 - d) \log(1 - X) \Big] , \qquad (3.16)$$

the minimization of (3.13) is to minimize the following **error function** between the teaching signal $d[i]$ and the output value of the machine $\langle d \rangle_{J,\mathbf{x}[i]}$ for the input data $\mathbf{x}[i]$,

$$L(\langle d \rangle_{J,\mathbf{x}[i]},\ d[i]) \qquad (3.17)$$

by adjusting J_x, J_y, J using the gradient method, etc., for each item of data $(\mathbf{x}[i], d[i])$. Since deep learning uses the stochastic gradient descent method, it is necessary to calculate the gradient of this error function with respect to J_x, J_y, and J. In this example, the gradient calculation is easy because it is a "one-layer"

[5] By the way, $-\sum P(\mathbf{x}, d) \log P(d|\mathbf{x})$ which is equivalent to the "J-independent part" in (3.13) is called conditional entropy. Due to the positive definite nature of the relative entropy, the first term of (3.13) should be greater than this conditional entropy. If (3.14), which approximates the first term, is smaller than this bound, it is clearly a sign of over-training. Since it is difficult to actually evaluate that term, we ignore the term in the following.

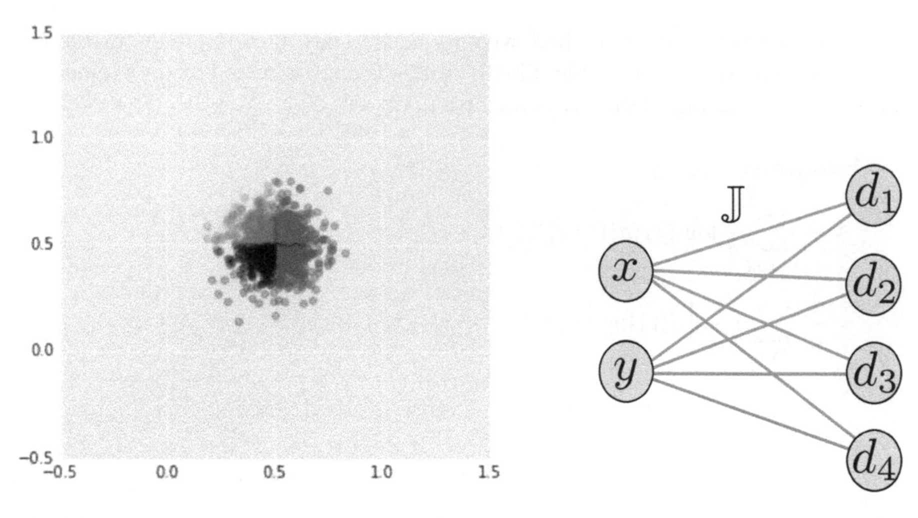

Fig. 3.2 Left : Data with (1, 0, 0, 0), blue; (0, 1, 0, 0), red; (0, 0, 1, 0), green; (0, 0, 0, 1), black. Right : Schematic diagram of the Hamiltonian (3.18)

system, but the deep neural network which will be introduced later has a slightly more complicated structure, so if there is no good gradient calculation method, calculation time will be consumed . Fortunately, there is an effective way to calculate gradients derived from the network structure. This will be explained later.

Multiclass classification
Suppose that the teaching signal is $\mathbf{d} = (d^1, d^2, d^3, d^4)$ with $\sum_{I=1}^{4} d^I = 1$, and let d^1 be blue, d^2 be red, d^3 be green, d^4 be black as shown in Fig. 3.2. Since d has increased to 4 components, the model Hamiltonian is modified to

$$H_{J,\mathbf{x}}(\mathbf{d}) = -\sum_{I=1}^{4}(xJ_{xI} + yJ_{yI} + J_I)d^I . \tag{3.18}$$

Since it is troublesome to write the sum symbol explicitly, we shall use a matrix notation: $\mathbb{J} = (J_{xI}, J_{yI})$ (4×2 matrix), $\mathbf{J} = (J_1, J_2, J_3, J_4)$ (4-dimensional vector), $\mathbf{x} = (x, y)$ (2-dimensional vector). Then,

$$H_{J,\mathbf{x}}(\mathbf{d}) = -\mathbf{d} \cdot (\mathbb{J}\mathbf{x} + \mathbf{J}) . \tag{3.19}$$

The Boltzmann weight with this Hamiltonian is

$$Q_J(\mathbf{d}|\mathbf{x}) = \frac{e^{-H_{J,\mathbf{x}}(\mathbf{d})}}{Z} = \frac{e^{\mathbf{d}\cdot(\mathbb{J}\mathbf{x}+\mathbf{J})}}{e^{(\mathbb{J}\mathbf{x}+\mathbf{J})_1} + e^{(\mathbb{J}\mathbf{x}+\mathbf{J})_2} + e^{(\mathbb{J}\mathbf{x}+\mathbf{J})_3} + e^{(\mathbb{J}\mathbf{x}+\mathbf{J})_4}} . \tag{3.20}$$

Here, let us introduce the following function called the **softmax function**, which is an extension of the sigmoid function,

$$\sigma_I(\mathbf{X}) = \frac{e^{X_I}}{\sum_J e^{X_J}} . \tag{3.21}$$

Using this softmax function, we can write the Boltzmann weight as

$$Q_J(d^I = 1|\mathbf{x}) = \sigma_I(\mathbb{J}\mathbf{x} + \mathbf{J}) . \tag{3.22}$$

As in the case with a single teaching signal, we define the model as

$$Q_J(\mathbf{x}, \mathbf{d}) = Q_J(d^I|\mathbf{x})P(\mathbf{x}). \tag{3.23}$$

Then we are to consider the minimization of the relative entropy

$$D_{KL}(P||Q_J) = -\sum_{\mathbf{x},\mathbf{d}} P(\mathbf{x}, d) \log Q_J(\mathbf{d}|\mathbf{x}) + (J\text{-independent part}) \tag{3.24}$$

and in particular,

$$\begin{aligned}\text{First term in (3.24)} &\approx -\sum_{i:\text{data}} \frac{1}{\#} \log Q_J(\mathbf{d}[i] \mid \mathbf{x}[i]) \\ &= \frac{-1}{\#} \sum_{i:\text{data}} \sum_{I=1}^{4} d_I[i] \log \sigma_I(\mathbb{J}\mathbf{x}[i] + \mathbf{J}) .\end{aligned} \tag{3.25}$$

As before, we look at the expectation value of the I component of $\mathbf{d}$,

$$\begin{aligned}\langle d_I \rangle_{J,\mathbf{x}[i]} &= \sum_{\mathbf{d}} d_I \cdot Q_J(\mathbf{d} \mid \mathbf{x}[i]) \\ &= Q_J(d_I = 1| \, \mathbf{x}[i]) = \sigma_I(\mathbb{J}\mathbf{x}[i] + \mathbf{J}) .\end{aligned} \tag{3.26}$$

So we introduce the cross entropy[6]

$$L(\mathbf{X}, \mathbf{d}) = -\sum_{I=1}^{4} d_I \log X_I , \tag{3.27}$$

[6]It is essentially the same as (3.16). To make them exactly the same, set $\mathbf{d} = (d, 1 - d)$ in (3.16).

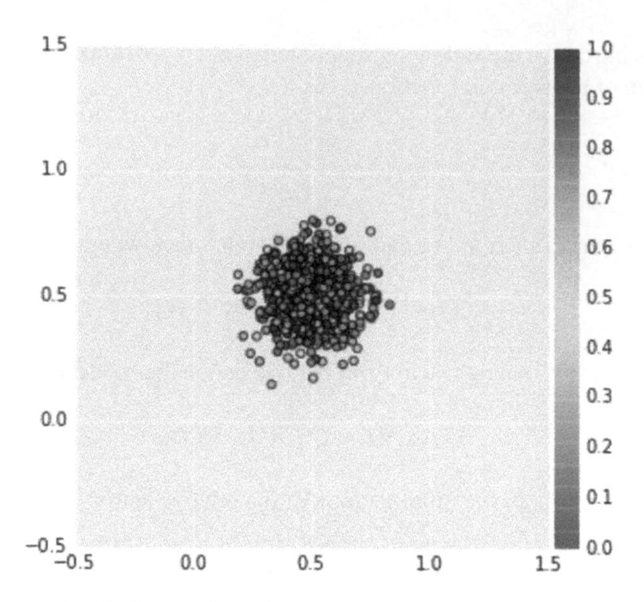

Fig. 3.3 The data with d being a real number

then after all, the minimization of (3.25) is a search for a matrix $\mathbb{J}$ and a vector $\mathbf{J}$ that reduce an error function

$$L(\langle \mathbf{d} \rangle_{J,\mathbf{x}[i]},\ \mathbf{d}[i]) \tag{3.28}$$

for each of the data.

Regression (Part 1)

In some cases, d may be a real number $d \in [-\infty, \infty]$, as shown in Fig. 3.3. For that case, if we take the same approach as above, we may define the Hamiltonian for real degrees of freedom,

$$H_{J,\mathbf{x}}(d) = \frac{1}{2}\Big(d - (\mathbb{J}\mathbf{x} + J)\Big)^2 . \tag{3.29}$$

Then we find a Gaussian distribution,

$$Q_J(d|\mathbf{x}) = \frac{e^{-\frac{1}{2}\left(d-(\mathbb{J}\mathbf{x}+J)\right)^2}}{Z} = \frac{e^{-\frac{1}{2}\left(d-(\mathbb{J}\mathbf{x}+J)\right)^2}}{\sqrt{2\pi}} . \tag{3.30}$$

The expectation value of d is the mean under the Gaussian distribution, so

$$\langle d \rangle_{J,\mathbf{x}[i]} = \mathbb{J}\mathbf{x}[i] + J . \tag{3.31}$$

And the error function is simply a logarithm of this, and turns out to be the mean square error,

$$L(\langle d \rangle_{J,\mathbf{x}[i]},\ d[i]) = \frac{1}{2}\Big(d[i] - \langle d \rangle_{J,\mathbf{x}[i]}\Big)^2 . \tag{3.32}$$

This is nothing but the **linear regression**.

Regression (Part 2)
For $d \in [0, \infty)$, a slightly more interesting effective modeling is known. Consider auxiliary N_{bits} degrees of freedom$\{h_{\text{bit}}^{(u)}\}_{u=1,2,\dots,N_{\text{bits}}}$ called rectified linear units (ReLU) [29] and prepare the following two Hamiltonians:

$$H_{J,\mathbf{x}}(\{h_{\text{bit}}^{(u)}\}) = -\sum_{u=1}^{N_{\text{bits}}} (\mathbb{J}\mathbf{x} + J + 0.5 - u) h_{\text{bit}}^{(u)} , \tag{3.33}$$

$$H_h(d) = \frac{1}{2}(d - h)^2 . \tag{3.34}$$

Here, (3.33) is the sum of (3.6) for all N_{bits} while shifting[7] J. And (3.34) is the Hamiltonian that gives the mean square error. We equate

$$h = \sum_{u=1}^{N_{\text{bits}}} h_{\text{bit}}^{(u)} , \tag{3.35}$$

while h is treated as an external field in (3.34). Then, from each Boltzmann distribution we find

$$Q_J(\{h_{\text{bit}}^{(u)}\}|\mathbf{x}) = \prod_{u=1}^{N_{\text{bits}}} \left\{ \begin{array}{ll} \sigma(\mathbb{J}\mathbf{x} + J + 0.5 - u) & (h_{\text{bit}}^{(u)} = 1) \\ 1 - \sigma(\mathbb{J}\mathbf{x} + J + 0.5 - u) & (h_{\text{bit}}^{(u)} = 0) \end{array} \right\} , \tag{3.36}$$

$$Q(d|h) = \frac{e^{-\frac{1}{2}(d-h)^2}}{\sqrt{2\pi}} . \tag{3.37}$$

From these two, we define the conditional probability

$$Q_J(d|\mathbf{x}) = \sum_{\{h_{\text{bit}}^{(u)}\}} Q\Big(d \Big| h = \sum_{u=1}^{N_{\text{bits}}} h_{\text{bit}}^{(u)}\Big) Q_J(\{h_{\text{bit}}^{(u)}\}|\mathbf{x}) . \tag{3.38}$$

[7]In this way, increasing the number of degrees of freedom while sharing the parameters corresponds to the handling of the ensemble of h_{bit} and improves the accuracy in a statistical sense. This is thought to lead to the improvement of performance [30]. In our case, J is shifted by 0.5, which simplifies $\langle h \rangle$ as shown in the main text, thus the computational cost is lower than that of the ordinary ensemble.

Then we find

$$
\begin{aligned}
&-\sum_{\mathbf{x},d} P(\mathbf{x}, d) \log Q_J(d|\mathbf{x}) \\
&\approx -\sum_{i:\text{Data}} \frac{1}{\text{The number of data}} \log Q_J(d[i]|\mathbf{x}[i]) \\
&= \frac{-1}{\text{The number of data}} \sum_{i:\text{Data}} \log\Big[\sum_{\{h_{\text{bit}}^{(u)}\}} Q\Big(d[i]\Big|h = \sum_{u=1}^{N_{\text{bits}}} h_{\text{bit}}^{(u)}\Big) Q_J(\{h_{\text{bit}}^{(u)}\}|\mathbf{x}[i])\Big]
\end{aligned}
\tag{3.39}
$$

Just as the probability mean for $\mathbf{x}$, d is replaced with the data (sample from P) mean, let also the probability mean for $\{h_{\text{bit}}^{(u)}\}$ in the log be replaced with the average under $Q_J(\{h_{\text{bit}}^{(u)}\}|\mathbf{x}[i])$. As a test, we calculate the average of h under $Q_J(\{h_{\text{bit}}^{(u)}\}|\mathbf{x}[i])$, then we find

$$
\langle h \rangle_{J,\mathbf{x}[i]} = \sum_{\{h_{\text{bit}}^{(u)}\}} \Big(\sum_{u=1}^{N_{\text{bits}}} h_{\text{bit}}^{(u)} \Big) Q_J(\{h_{\text{bit}}^{(u)}\}|\mathbf{x}[i]) = \sum_{u=1}^{N_{\text{bits}}} \sigma(\mathbb{J}\mathbf{x} + J + 0.5 - u)\,. \tag{3.40}
$$

Defining $z = \mathbb{J}\mathbf{x} + J$ and plotting this with z as the horizontal axis, we find the blue line in Fig. 3.4.

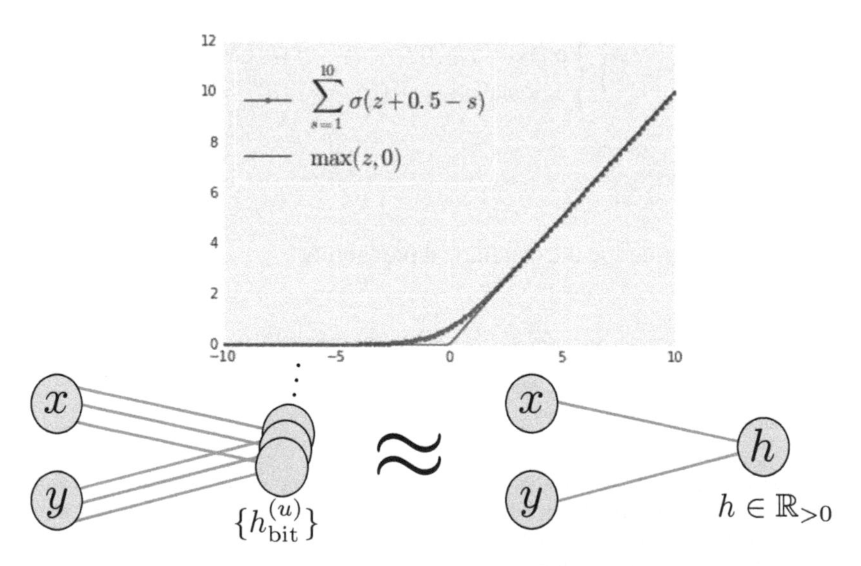

Fig. 3.4 Top: Example of $N_{\text{bits}} = 10$. The limit $N_{\text{bits}} \to \infty$ looks similar to $\max(z, 0)$. Bottom: Schematic diagram of the ReLU

As can be seen from the figure, the expectation value of h gradually approaches $\max(\mathbb{J}\mathbf{x} + J, 0)$ as the number of bits increases. Let us call it $\sigma_{\text{ReLU}}(z) = \max(z, 0)$, and if we adopt the approximation, we find:

$$\langle h \rangle_{J,\mathbf{x}[i]} \approx \sigma_{\text{ReLU}}(\mathbb{J}\mathbf{x}[i] + J)\,, \tag{3.41}$$

$$\sum_{\{h_{\text{bit}}^{(u)}\}} Q\Big(d[i]\Big|h = \sum_{u=1}^{N_{\text{bits}}} h_{\text{bit}}^{(u)}\Big) Q_J(\{h_{\text{bit}}^{(u)}\}|\mathbf{x}[i]) \approx Q\Big(d[i]\Big|h = \langle h \rangle_{J,\mathbf{x}[i]}\Big)\,. \tag{3.42}$$

Furthermore, the error function is

$$-\log Q\Big(d[i]\Big|h = \langle h \rangle_{J,\mathbf{x}[i]}\Big) \approx \frac{1}{2}\Big(d[i] - \sigma_{\text{ReLU}}(\mathbb{J}\mathbf{x}[i] + J)\Big)^2\,. \tag{3.43}$$

Multi-component teaching signal
In addition to the multi-component labels as shown in Fig. 3.2, we can set $d^I \in \{0, 1\}$ or $d^I \in \mathbb{R}$. In that case, we need to consider the output of applying σ or σ_{ReLU} to each component.

3.1.2 Deep Neural Network

Extending the two-Hamiltonian approach, like the ReLU above, naturally leads to the idea of a **deep model**. A deep neural network is a neural network that has multiple intermediate layers between inputs and outputs, as shown in Fig. 3.5.

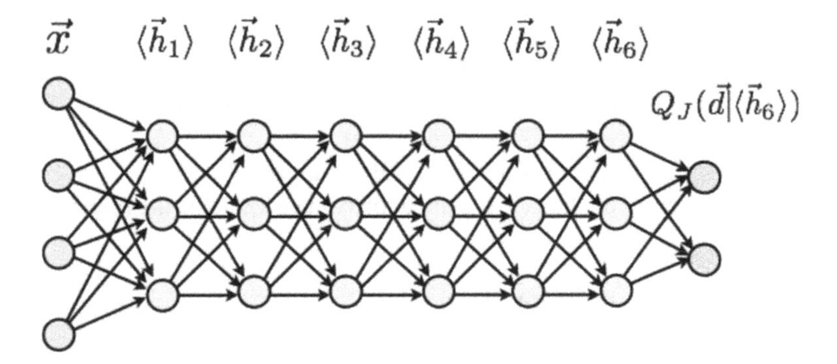

Fig. 3.5 Deep neural network. The output value obtained by applying the nonlinear function $\sigma_\bullet$ of each layer corresponds to an expectation value from a statistical mechanics standpoint

$(N-1)$ extra degrees of freedom are prepared, and considering the corresponding N Hamiltonians

$$H^1_{J_1,\mathbf{x}}(\mathbf{h}_1)\,, \tag{3.44}$$

$$H^2_{J_2,\mathbf{h}_1}(\mathbf{h}_2)\,, \tag{3.45}$$

$$\dots \tag{3.46}$$

$$H^N_{J_N,\mathbf{h}_{N-1}}(\mathbf{d})\,, \tag{3.47}$$

the model is

$$Q_J(\mathbf{d}|\mathbf{x}) = \sum_{\mathbf{h}_1,\mathbf{h}_2,\dots,\mathbf{h}_{N-1}} Q_{J_N}(\mathbf{d}|\mathbf{h}_{N-1})\dots Q_{J_2}(\mathbf{h}_2|\mathbf{h}_1)Q_{J_1}(\mathbf{h}_1|\mathbf{x})\,. \tag{3.48}$$

And we adopt the approximation that replaces the sum with the average,

$$Q_J(\mathbf{d}|\mathbf{x}) \approx Q_{J_N}(\mathbf{d}|\langle\mathbf{h}_{N-1}\rangle)\,, \quad \langle\mathbf{h}_{N-1}\rangle = \sigma_{N-1}(\mathbb{J}_{N-1}\langle\mathbf{h}_{N-2}\rangle + \mathbf{J}_{N-1})\,, \tag{3.49}$$

$$\langle\mathbf{h}_{N-2}\rangle = \sigma_{N-2}(\mathbb{J}_{N-2}\langle\mathbf{h}_{N-3}\rangle + \mathbf{J}_{N-2})\,, \tag{3.50}$$

$$\dots$$

$$\langle\mathbf{h}_1\rangle = \sigma_1(\mathbb{J}_1\mathbf{x} + \mathbf{J}_1)\,. \tag{3.51}$$

Here $\sigma_\bullet$ is a function determined from the corresponding Hamiltonian (such as σ or σ_{ReLU} above), called the **activation function**.

You can see that the structure repeats the following:

1. Linear transformation by $\mathbb{J}$, $\mathbf{J}$
2. Nonlinear transformation by activation function

Each element of the repeating structure is called a **layer**.

Furthermore, for given data $(\mathbf{x}, \mathbf{d})$, the approximate value of the relative entropy is written as

$$-\log Q_{JN}(\mathbf{d}|\langle\mathbf{h}_{N-1}\rangle) = L(\mathbf{d}, \langle\mathbf{h}_N\rangle), \quad \langle\mathbf{h}_N\rangle = \sigma_N(\mathbb{J}_N\langle\mathbf{h}_N\rangle + \mathbf{J}_N)\,. \tag{3.52}$$

Omitting the part in $\mathbf{J}$ for brevity, this means calculating the difference between $\mathbf{d}$ and

$$\langle\mathbf{h}_N\rangle = \sigma_N\Big(\mathbb{J}_N\sigma_{N-1}\Big(\dots\mathbb{J}_2\sigma_1(\mathbb{J}_1\mathbf{x})\dots\Big)\Big)\,. \tag{3.53}$$

In the usual context of deep learning, the parameters are often written as

$$\mathbb{J} = W, \tag{3.54}$$

$$\mathbf{J} = \mathbf{b}, \tag{3.55}$$

and referred to as weight W and bias $\mathbf{b}$.

The neural network has been "derived." In the next section, we will look at how learning can proceed.

3.2 Derivation of Backpropagation Method Using Bracket Notation

We shall derive the famous **backpropagation method** of the neural network [31]. The backpropagation method is just a combination of the differential method of composite functions and a little bit of linear algebra. Here we derive it by using the quantum mechanics notation representing **ket** $|\bullet\rangle$ as an element in a vector space and **bra** $\langle\bullet|$ as its dual. We define

$$\langle \mathbf{h}_l \rangle := \begin{pmatrix} \langle h_l^1 \rangle \\ \langle h_l^2 \rangle \\ \dots \\ \langle h_l^{n_l} \rangle \end{pmatrix} = |h_l\rangle \,, \tag{3.56}$$

and the basis vector[8]

$$|m\rangle = \begin{pmatrix} 0 \\ \vdots \\ 0 \\ 1 \\ 0 \\ \vdots \\ 0 \end{pmatrix} \leftarrow m\text{th}. \tag{3.57}$$

For the sake of simplicity, we ignore the $\mathbf{J}$ part for now; then the equation

$$\langle \mathbf{h}_l \rangle = \sigma_l(\mathbb{J}_l \langle \mathbf{h}_{l-1} \rangle) \quad \text{(Apply activation function to each component)} \tag{3.58}$$

[8] Here, every $|h_l\rangle$ is expanded by the basis $|m\rangle$, meaning that the dimensions of all vectors are the same. However, if the range in the sum symbol of (3.59) is not abbreviated, this notation can be applied in different dimensions.

can be written as

$$|h_l\rangle = \sum_m |m\rangle \sigma_l(\langle m|\mathbb{J}_l|h_{l-1}\rangle)\,. \tag{3.59}$$

Furthermore, the "variation" of (3.59) with respect to the parameter $\mathbb{J}$ is calculated as

$$\begin{aligned}\delta|h_l\rangle &= \underbrace{\sum_m |m\rangle \sigma_l'(\langle m|\mathbb{J}_l|h_{l-1}\rangle)\langle m|}_{\text{We name this part } \mathbb{G}_l} \Big(\delta\mathbb{J}_l|h_{l-1}\rangle + \mathbb{J}_l\delta|h_{l-1}\rangle\Big) \\ &= \mathbb{G}_l\Big(\delta\mathbb{J}_l|h_{l-1}\rangle + \mathbb{J}_l\delta|h_{l-1}\rangle\Big)\,. \end{aligned} \tag{3.60}$$

Also, using the notation

$$\mathbf{d} = |d\rangle\,, \tag{3.61}$$

with the activation function of the final layer as $\sigma_N(\langle m|z\rangle) = -\log\sigma_m(\mathbf{z})$, the softmax cross entropy that appears in the error function is

$$E = \langle d|h_N\rangle\,. \tag{3.62}$$

If we want to consider the mean square error, σ_N can be anything, and we can write

$$E = \frac{1}{2}(|d\rangle - |h_N\rangle)^2 = \frac{1}{2}\langle h_N - d|h_N - d\rangle\,. \tag{3.63}$$

In any case, the variation on the output of the final layer is written with the bra vector,

$$\frac{\delta E}{\delta|h_N\rangle} = \begin{cases}\langle d| \\ \langle h_N - d|\end{cases} =: \langle\delta_0|. \tag{3.64}$$

We have named this $\langle\delta_0|$. If we take the variation of the error function for all $\mathbb{J}_l$ instead of the final layer output, δE can be transformed as follows by using (3.60) repeatedly:

$$\begin{aligned}\delta E &= \frac{\delta E}{\delta|h_N\rangle}\delta|h_N\rangle = \langle\delta_0| \underbrace{\delta|h_N\rangle}_{\text{Transform this with (3.60)}} \\ &= \underbrace{\langle\delta_0|\mathbb{G}_N}_{\text{Define this }=:\langle\delta_1|} \Big(\delta\mathbb{J}_N|h_{N-1}\rangle + \mathbb{J}_N\delta|h_{N-1}\rangle\Big)\end{aligned}$$

$$= \langle\delta_1|\delta\mathbb{J}_N|h_{N-1}\rangle + \langle\delta_1|\mathbb{J}_N \underbrace{\delta|h_{N-1}\rangle}_{\text{Transform this with (3.60)}}$$

$$= \langle\delta_1|\delta\mathbb{J}_N|h_{N-1}\rangle + \underbrace{\langle\delta_1|\mathbb{J}_N\mathbb{G}_{N-1}}_{\text{Define this }=:\langle\delta_2|} \Big(\delta\mathbb{J}_{N-1}|h_{N-2}\rangle + \mathbb{J}_{N-1}\delta|h_{N-2}\rangle\Big)$$

$$= \cdots$$

$$= \langle\delta_1|\delta\mathbb{J}_N|h_{N-1}\rangle + \langle\delta_2|\delta\mathbb{J}_{N-1}|h_{N-2}\rangle + \cdots + \langle\delta_N|\delta\mathbb{J}_1|h_0\rangle\,. \tag{3.65}$$

Here the last $|h_0\rangle = \sum_m |m\rangle\langle m|x\rangle = \sum_m |m\rangle x^m$ is a ket representation of $\mathbf{x}$. Therefore, if we want to reduce the value of the error function, we just take

$$\delta\mathbb{J}_l = -\epsilon|\delta_{N-l+1}\rangle\langle h_{l-1}|\,. \tag{3.66}$$

This is because with this we find

$$\delta E = -\epsilon\Big(\big||\delta_1\rangle\big|^2\big||h_{N-1}\rangle\big|^2 + \big||\delta_2\rangle\big|^2\big||h_{N-2}\rangle\big|^2 + \ldots\big||\delta_N\rangle\big|^2\big||h_0\rangle\big|^2\Big)\,, \tag{3.67}$$

and the change of E becomes a small negative number. We can repeat this with various pairs (x, d). By the way, the right-hand side of (3.66) is minus ϵ times the differentiation of the error function E, so this is a stochastic gradient descent method described in the previous chapter. Bra vector $\langle\delta_l|$ artificially introduced in the derivation of (3.65) satisfies the recurrence formula

$$\langle\delta_l| = \langle\delta_{l-1}|\mathbb{J}_{N-l+2}\mathbb{G}_{N-l+1}\,, \tag{3.68}$$

with $\mathbb{J}_{N+1} = 1$, which is an expression "similar" to (3.59). The backpropagation algorithm is an algorithm that calculates the differential value of E for all $\mathbb{J}_l$ by combining (3.59), (3.66) and (3.68). It is

1. Repeat (3.59) with an initial value $|h_0\rangle = |x[i]\rangle$, and calculate $|h_l\rangle$.
2. Repeat (3.68) with an initial value $\langle\delta_0| = \langle d[i]|$, and calculate $\langle\delta_l|$.
3. $\nabla_{\mathbb{J}_l} E = |\delta_{N-l+1}\rangle\langle h_{l-1}|$. (3.69)

The name "backpropagation" means that the propagation direction of $\langle\delta|$ by (3.68) is opposite to the direction of (3.59). This fact is written in our bracket notation as

$$|h\rangle \text{ : ket : Forward propagation,} \tag{3.70}$$

$$\langle\delta| \text{ : bra : Backpropagation.} \tag{3.71}$$

The correspondence is visually clear.

3.3 Universal Approximation Theorem of Neural Network

Why can a neural network express the connection (correlation) between data? The answer, in a certain limit, is the **universal approximation theorem** of neural networks [32]. Similar theorems have been proved by many people, and here we will explain a simple proof by M. Nielsen [33].[9]

Universal approximation theorem of one-dimensional neural network
Let us take the simplest model first. Consider a neural network with a single hidden layer, that gives a one-dimensional real number when given a one-dimensional variable x. It looks like this:

$$f(x) = \mathbf{J}^{(2)} \cdot \sigma_{\text{step}}(\mathbf{J}^{(1)} x + \mathbf{b}^{(1)}) + \mathbf{b}^{(2)} . \tag{3.72}$$

$\sigma_{\text{step}}(x)$ is **step function** which serves as an activation function,

$$\sigma_{\text{step}}(x) = \begin{cases} 0 & (x < 0) , \\ 1 & (x \geq 0) . \end{cases} \tag{3.73}$$

When the activation function is a sigmoid function, it can be obtained at some limit of it.[10] When the argument is multi-component, we define that the operation is on each component. $\mathbf{J}^{(i)}$ is the weight , and $\mathbf{b}^{(i)}$ is the bias.

$$\mathbf{J}^{(l)} = (j_1^{(l)}, j_2^{(l)}, j_3^{(l)}, \cdots , j_{n_{\text{unit}}}^{(l)})^\top , \tag{3.74}$$

$$\mathbf{b}^{(l)} = (b_1^{(l)}, b_2^{(l)}, b_3^{(l)}, \cdots , b_{n_{\text{unit}}}^{(l)})^\top . \tag{3.75}$$

Here $j_i^{(l)}$ and $\mathbf{b}^{(l)}$ are real numbers. l is the serial number of the layer, $l = 1, 2$. n_{unit} is the number of units in each layer, which will be given later.

What happens at the first and second layers
First, let us see what happens at the first and second layers with $n_{\text{unit}} = 1$. In this case, the following is input to the third layer:

$$g_1(x) = \sigma_{\text{step}}(j_1^{(1)} x + b_1^{(1)}) . \tag{3.76}$$

If we vary $j_1^{(1)}$ and $b_1^{(1)}$, we find that the step function moves left or right. Then multiply this by the coefficient $j_1^{(2)}$,

$$g_2(x) = j_1^{(2)} \sigma_{\text{step}}(j_1^{(1)} x + b_1^{(1)}) . \tag{3.77}$$

[9]Visit his website for an intuitive understanding of the proof.

[10]The situation is the same as the physics calculations at zero temperature at which the Fermi distribution function becomes a step function.

Then we can freely change the height (even in the negative direction). Adding a bias term $b_1^{(2)}$ for the second layer can also change the minimum value of the step function.

Then, let $n_{\text{unit}} = 2$. We can draw a rectangle function of any shape using $\mathbf{J}^{(l)}$ and $\mathbf{b}^{(l)}$ as parameters. Now we are ready for the proof. Consider $n_{\text{unit}} = 2k$ $(k > 1)$. Then we can draw a graph with many rectangle functions superimposed.

Given a continuous objective function $t(x)$, the neural network $f(x)$ considered above can approximate it by increasing n_{unit} and adjusting the parameters. One can get as close as one likes.[11] Therefore, a one-dimensional neural network can represent any continuous function.

Universal approximation theorem of neural networks
The facts mentioned above can be extended to higher dimensions. That is, a continuous mapping from $\mathbf{x}$ to $\mathbf{f}(x)$ can be expressed using a neural network. This fact is called the universal approximation theorem of neural networks.

At the same time, note that this theorem refers to an unrealistic situation. First, the activation function is a step function. In the learning based on the backpropagation method, the activation function is required to be differentiable, so the step function cannot meet the purpose. Second, we need an infinite number of intermediate units. As a quantitative problem, it does not describe how the neural network improves its approximation to the target function at what speed and with what accuracy. Another caveat is that this theorem only says that a neural network can give an approximation of the desired function. Nevertheless, this theorem, to some extent, intuitively explains a fragment of why neural networks are powerful.

Why deep layers?
So far, we have seen a neural network with a single hidden layer. Even if the number of hidden layers is one, if the number of units is infinite, any function can be approximated. Then, why is deep learning effective?

The following facts are known [34]:

1. More units in the middle layer $\Rightarrow$ Expressivity increases in power
2. Deeper neural network $\Rightarrow$ Expressivity increases exponentially

In this section, we will consider why deepening is useful, using a toy model.

Toy model of deep learning
Let us introduce a simple toy model that can show that the number of hidden layers affects the expressivity of the neural network. A neural network with one hidden layer is

$$f_{\text{1h-NN}}(x) = \mathbf{j}_L \cdot \sigma_{\text{act}}(\mathbf{j}_0 x + \mathbf{b}_0) + b_L\,, \tag{3.78}$$

[11] The part corresponds to the dense nature in Cybenko's proof.

with $\dim[\mathbf{b}_0] = N_{\text{unit}}$. Here assume that the activation function is not an even function. A neural network with two hidden layers is

$$f_{\text{2h-NN}}(x) = \mathbf{j}_L \cdot \sigma_{\text{act}}(J\sigma_{\text{act}}(\mathbf{j}_0 x + \mathbf{b}_0) + \mathbf{b}_1) + b_L . \tag{3.79}$$

Let us compare these and look at the meaning of deepening.

The case with a linear activation function

As a preparation, let us consider the case where the activation function is linear, especially an identity map. That is,

$$\sigma_{\text{act}}(x) = x . \tag{3.80}$$

In this case, for a single hidden layer, we find a linear map,

$$f_{\text{1h-NN}}(x) = \mathbf{j}_L \cdot (\mathbf{j}_0 x + \mathbf{b}_0) + b_L \tag{3.81}$$

$$= (\mathbf{j}_L \cdot \mathbf{j}_0)x + (\mathbf{j}_L \cdot \mathbf{b}_0 + b_L) . \tag{3.82}$$

Furthermore, even with two layers, we find

$$f_{\text{2h-NN}}(x) = \mathbf{j}_L \cdot (J_1(\mathbf{j}_0 x + \mathbf{b}_0) + \mathbf{b}_1) + b_L \tag{3.83}$$

$$= (\mathbf{j}_L \cdot J_1 \mathbf{j}_0)x + (\mathbf{j}_L \cdot J_1 \mathbf{b}_0 + \mathbf{j}_L \cdot \mathbf{b}_1 + b_L) . \tag{3.84}$$

In other words, if we make the activation function an identity map, we will always get only a linear map.

The case with a nonlinear activation function

Next, consider the case where the activation function is nonlinear. In particular, we want to consider the power as the complexity of a function that can be expressed by a neural network, so we take

$$\sigma_{\text{act}}(x) = x^3 . \tag{3.85}$$

For the sake of simplicity, assume that the neural network has a single hidden layer with two units, $\mathbf{j}_0 = (j_1^0,\ j_2^0)^\top, \mathbf{j}_L = (j_1^L,\ j_2^L)^\top$. Then we find

$$f_{\text{1h-NN}}(x) = \mathbf{j}_L \cdot \sigma_{\text{act}}((j_1^0 x,\ j_2^0 x)^\top + (b_1^0,\ b_2^0)^\top) + b_L \tag{3.86}$$

$$= \mathbf{j}_L \cdot \sigma_{\text{act}}((j_1^0 x + b_1^0,\ j_2^0 x + b_2^0)^\top) + b_L \tag{3.87}$$

$$= \mathbf{j}_L ((j_1^0 x + b_1^0)^3,\ (j_2^0 x + b_2^0)^3)^\top + b_L \tag{3.88}$$

$$= j_1^L (j_1^0 x + b_1^0)^3 + j_2^L (j_2^0 x + b_2^0)^3 + b_L . \tag{3.89}$$

So, this neural network can express up to a cubic function, with seven parameters. In the case of a general number of units, fitting is performed by a cubic function using $3N_{\text{unit}} + 1$ parameters.

Conversely, considering the case where the number of units is 1 but with two hidden layers, we find a ninth-order function,

$$f_{\text{2h-NN}}(x) = \mathbf{j}_L \cdot \sigma_{\text{act}}(J\sigma_{\text{act}}(\mathbf{j}_0 x + \mathbf{b}_0) + \mathbf{b}_1) + b_L \tag{3.90}$$

$$= j_L(j(j_0 x + b_0)^3 + b_1)^3 + b_L\,. \tag{3.91}$$

The number of parameters is 6. In general, when the order of the activation function is n and the number of hidden layers is h, the maximum order of the expressed function is $x^{(h+1)n}$. We can see that all orders appear when the activation function is an odd function.

According to this model, the complexity of the functions that can be expressed does not change even if the number of units in the middle layer is increased with just three layers. On the other hand, if we increase the number of layers with an activation function with $n > 1$, the complexity of the function causes a combinatorial explosion. In other words, more complex functions can be represented by neural networks.

Finally, it should be emphasized that the discussion here is only with a model for intuitive understanding. In particular, in the case of the activation function ReLU, which is very often used, the argument here does not hold because of its linearity. Nevertheless, since ReLU is also subject to the universal approximation [35] in fact, it can be said that deep learning can be used with confidence by trusting the universal approximation.

Column: Statistical Mechanics and Quantum Mechanics

Canonical Distribution in Statistical Mechanics

The statistical mechanics used in this book is called **canonical distribution** for a given temperature. The physical situation is as follows. Suppose we have some physical degrees of freedom, such as a spin or a particle position. Denote it as d. Since this is a physical degree of freedom, it should have energy which depends on its value. Let it be $H(d)$. When this system is immersed in an environment with a temperature of T (called a heat-bath), the probability of achieving d with the energy $H(d)$ is known to be

$$P(d) = \frac{e^{-\frac{H(d)}{k_B T}}}{Z}\,. \tag{3.92}$$

Here, k_B is an important constant called the **Boltzmann constant**, and Z is called the **partition function**, defined as

$$Z = \sum_d e^{-\frac{H(d)}{k_B T}} . \tag{3.93}$$

In the main text, we put $k_B T = 1$.

Simple example: law of equipartition of energy
As an example, consider d as the position x and momentum p of a particle in a box of length L, with the standard kinetic energy function as its energy. The expectation value of energy is

$$\langle E \rangle = \int_0^L dx \int_{-\infty}^{+\infty} dp \, \frac{p^2}{2m} \frac{e^{-\frac{p^2}{2mk_B T}}}{Z} . \tag{3.94}$$

Here

$$Z = \int_0^L dx \int_{-\infty}^{+\infty} dp \, e^{-\frac{p^2}{2mkT}} = L(2\pi m k_B T)^{1/2} . \tag{3.95}$$

The calculation of $\langle E \rangle$ looks a bit difficult, but using $\beta = \frac{1}{k_B T}$ we find

$$Z = L\left(\frac{2\pi m}{\beta}\right)^{1/2} , \tag{3.96}$$

$$\langle E \rangle = -\frac{\partial}{\partial \beta} \log Z = \frac{1}{2}\frac{1}{\beta} = \frac{1}{2} k_B T . \tag{3.97}$$

In other words, when the system temperature T is high (i.e. when the system is hot), the expectation value of energy is high, and when the temperature is low, the expectation value of energy is low. This is consistent with our intuition. In addition, if we consider the case of three spatial dimensions, we can obtain the famous formula $\frac{3}{2} k_B T$ as the expectation value of energy.

Bracket Notation in Quantum Mechanics

In the derivation of the backpropagation method, **bracket notation** has been introduced as a simple method. This is nothing more than just writing a vector as a **ket**,

$$\mathbf{v} = |v\rangle . \tag{3.98}$$

There are some notational conveniences. For example, the inner bracket is written conventionally as

$$\mathbf{w} \cdot \mathbf{v} = \langle w|v \rangle, \tag{3.99}$$

which means that the inner product is regarded as the "product of the matrices,"

$$\mathbf{w} \cdot \mathbf{v} = \begin{pmatrix} w_1, & w_2, & \cdots \end{pmatrix} \begin{pmatrix} v_1 \\ v_2 \\ \vdots \end{pmatrix} . \tag{3.100}$$

Also in the main text we have had

$$|v\rangle\langle w| \tag{3.101}$$

and also this can be regarded as a "matrix," made by a matrix product,

$$|v\rangle\langle w| = \begin{pmatrix} v_1 \\ v_2 \\ \vdots \end{pmatrix} \begin{pmatrix} w_1, & w_2, & \cdots \end{pmatrix} = \begin{pmatrix} v_1 w_1 & v_1 w_2 & \cdots \\ v_2 w_1 & v_2 w_2 & \cdots \\ \cdots & & \end{pmatrix} . \tag{3.102}$$

Chapter 4
Advanced Neural Networks

Abstract In this chapter, we explain the structure of the two types of neural networks that have been the mainstays of deep learning in recent years, following the words of physics in the previous chapter. A convolutional neural network has a structure that emphasizes the spatial proximity in input data. Also, recurrent neural networks have a structure to learn input data in time series. You will learn how to provide a network structure that respects the characteristics of data.

In Chap. 3, using physics language, we have described the introduction of the most basic neural network (forward propagation neural network) and the associated error function and the gradient calculation of the empirical error (error backpropagation method). This chapter is for the following two representative examples of a recent deep learning architecture:

- Convolutional neural network
- Recurrent neural network

and we will explain the structure of these and related topics.

4.1 Convolutional Neural Network

4.1.1 Convolution

Consider a two-dimensional image. For a grayscale image, the input value is x_{ij} as the ijth pixel value, and for a color image, it is x_{ij}^c as the ijth pixel value, with c for the cth color channel. So we naturally consider the tensor structure. As an example, let us consider d_{IJ} as

- Probability that a cat is shown around $i \approx I, j \approx J$. (4.1)

A. Tanaka et al., *Deep Learning and Physics*, Mathematical Physics Studies,
https://doi.org/10.1007/978-981-33-6108-9_4

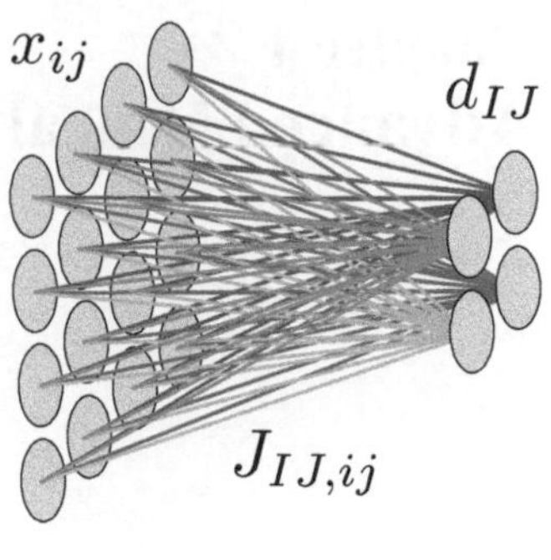

Fig. 4.1 Schematic diagram of the coupling constant of the Hamiltonian (4.2). Color in lines corresponds to each IJ

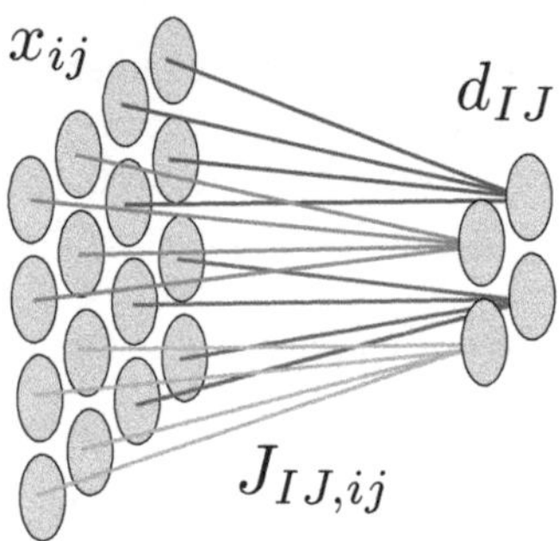

Fig. 4.2 Schematic diagram of the coupling constant (4.4). Color in lines corresponds to each IJ

In this case, the simplest Hamiltonian is

$$H_{J,x}(\{d_{IJ}\}) = -\sum_{IJ}\Big(\sum_{ij} d_{IJ}J_{IJ,ij}x_{ij} + d_{IJ}J_{IJ}\Big). \tag{4.2}$$

As shown in Fig. 4.1. However, for discussing (4.1), it would be useless to consider all ij and all interactions. Furthermore, when considering (4.1), it is better to define how far away from ij we have to take into account. Therefore we want to take

$$J_{IJ,ij} = \begin{cases} \text{non-zero} \begin{pmatrix} i = s_1 I + \alpha, & \alpha \in [-W_1/2, W_1/2] \\ j = s_2 J + \beta, & \beta \in [-W_2/2, W_2/2] \end{pmatrix} \\ \text{zero} \qquad \text{others.} \end{cases} \tag{4.3}$$

Here, s_1, s_2 are natural numbers called **strides**, which are the parameters for how many pixels to skip each time to grasp the feature. W_1, W_2 are natural numbers called **filter sizes**, and are parameters for how large the area is to capture the feature. This is realized by restricting the **coupling constant** to (see Fig. 4.2)

$$J_{IJ,ij} = \sum_{\alpha\beta} J_{IJ,\alpha\beta}\delta_{i,s_1 I+\alpha}\delta_{j,s_2 J+\beta}\,. \tag{4.4}$$

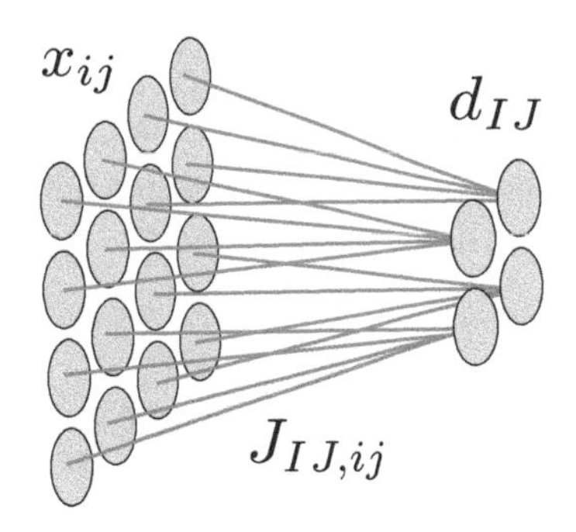

Fig. 4.3 Schematic diagram of the coupling constant (4.5). The variables here are only the degrees of freedom $J_{\alpha\beta}$

Also, if the feature of the image we want to capture is "cat-like," that feature should not depend on the pixel coordinates I, J of the image. This is because sometimes a cat appears in the upper right corner of the photo, and sometimes in the middle. Then, it is natural for the purpose of (4.1) not to include the IJ dependence in the coupling constant. This reduces the coupling constant to the form

$$J_{IJ,ij} = \sum_{\alpha\beta} J_{\alpha\beta} \delta_{i,s_1 I+\alpha} \delta_{j,s_2 J+\beta} \,. \tag{4.5}$$

See Fig. 4.3. Then, it leads to a Hamiltonian,

$$H_{J,x}^{\text{conv}}(\{d_{IJ}\}) = -\sum_{IJ} d_{IJ} \Big(\sum_{\alpha\beta} J_{\alpha\beta} x_{s_1 I+\alpha, s_2 J+\beta} + J \Big). \tag{4.6}$$

Calculating the Boltzmann weights with this Hamiltonian, we find

$$Q_J(\{d_{IJ} = 1\}|x) = \sigma \Big(\sum_{\alpha\beta} J_{\alpha\beta} x_{s_1 I+\alpha, s_2 J+\beta} + J \Big). \tag{4.7}$$

The operation

$$x_{ij} \to \sum_{\alpha\beta} J_{\alpha\beta} x_{s_1 I+\alpha, s_2 J+\beta} \tag{4.8}$$

is called a convolution,[1] and a neural network that includes a convolution operation is called a convolutional neural network. It was first introduced in [36].[2] As

[1]This convolution is essentially the same as that used in mathematical physics, although the signs may be different.

[2]It was written in [36] that the author got the idea of the convolutional neural network from the paper by Hubel and Wiesel [37] which pointed out that there are two types of cells in the visual cortex, and from the paper by Fukushima and Miyake which implemented those to a neural network [38].

one of the origins of the recent deep learning boom, it is often cited that a convolutional neural network (AlexNet [39][3]) in an image recognition competition using ImageNet [26] achieved a truly "mind-boggling" performance in 2012. Since then, convolutional neural networks in **image recognition** have become as much the norm as "sushi is tuna."[4]

4.1.2 Transposed Convolution

By the way, if we consider the convolution as the "interaction" between the image x_{ij} and the feature d_{IJ} as described above, it is unnatural to pay attention to only the convolution operation (that produces something equivalent to d_{IJ} from the input x_{ij}). Namely, it is natural to consider an operation whose input is the feature d_{IJ} and output is the image x_{ij}. This is considered to be just an operation by a transposed matrix, if we look at the Hamiltonian (4.6) as a quadratic form of $\mathbf{x}$ and $\mathbf{d}$ by appropriately changing the indices. This transposed convolution

$$d_{IJ} \to \sum_{IJ} d_{IJ} J_{IJ,ij} \tag{4.9}$$

would be used, for example, when we want to generate a cat-like image at ij by an input of "cat-likeness." See Fig. 4.4. Readers may have heard of the news that a picture drawn by an artificial intelligence was sold at an auction at a high price. This transposed convolution is a technique often used to implement neural networks related to such image generation.[5] As an example, Fig. 4.5 shows the result of a neural network called **DCGAN** (deep convolutional generative adversarial network) [42], about an unsupervised learning of handwritten characters.[6] The neural network is given only the input image of **MNIST**, and by learning its features (stopping, hitting and flipping, etc.) well, it generates an image that mimics MNIST. We can see how successful it is.

Checkerboard artifact
Transpose convolution is considered to be a natural method of generating high-dimensional features from low-dimensional features as an inverse operation of

---

[3]Alex is the name of the first author of this paper.

[4]Speaking of tuna in the past, "pickling" was used to prevent rot, so it was common to eat lean meat, and the toro part seemed to be worthless. However, thanks to the development of refrigeration technology, fresh tuna can be delivered to areas far away from the sea, which has led to the spread of toro. Similarly, if new technologies develop in machine learning, a "next-generation" architecture, possibly beyond convolutional neural networks, may emerge. In fact, in recent years a new neural network called Capsule Network [40] has attracted attention.

[5]For readers who want to learn more, we suggest referring to [41], for example.

[6]We will explain this method in detail, in Sect. 6.3.

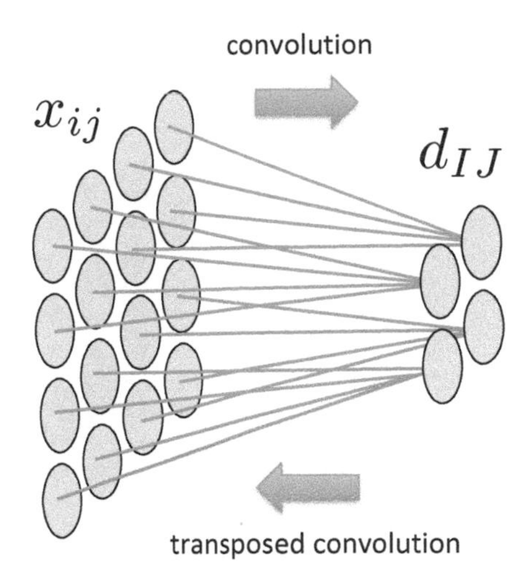

Fig. 4.4 Relation between the convolution and the **transposed convolution**

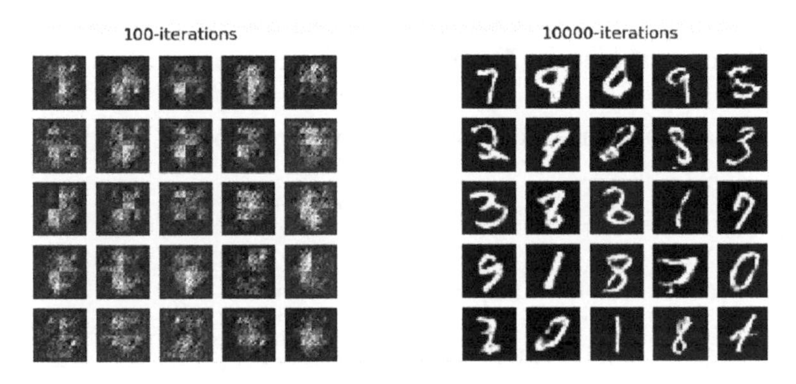

Fig. 4.5 Handwritten characters generated by a DCGAN. Left: Checkerboard pattern is noticeable at the beginning of learning. Right: Later in learning, it reached a level at which the pattern is not noticeable

convolution. However, due to the nature of the definition, a unique pattern called **checkerboard artifact** is often generated (see Figs. 4.5 left and 4.6).

In a deep neural network, if the training up to the input layer d_{IJ} has not progressed—for example, if d_{IJ} becomes almost uniform—then the transposed convolution will simply be just shifting and adding the filters in the calculation, resulting in the checkerboard artifact. If the training of the input layer d_{IJ} progresses well, the last d_{IJ} will not be uniform and the checkerboard pattern will not be visible.

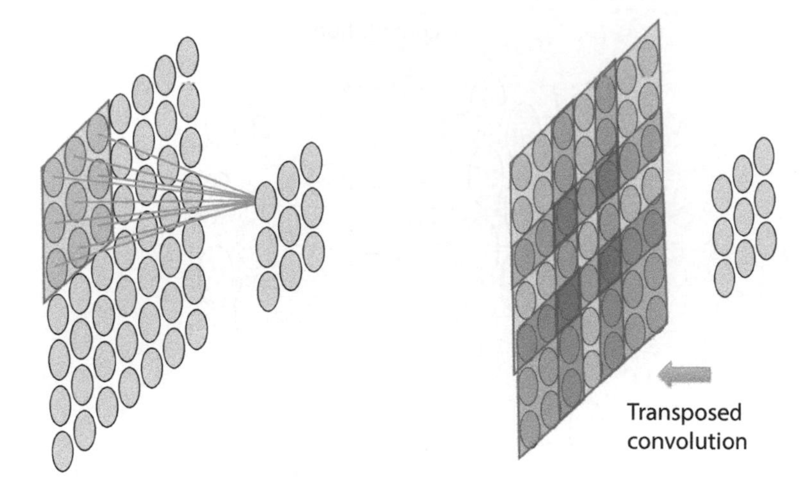

Fig. 4.6 An intuitive description that can create checkerboard artifacts

4.2 Recurrent Neural Network and Backpropagation

So far, we have considered the following one-way propagation:

$$|h\rangle = \mathbb{T}|x\rangle := \sum_m |m\rangle \sigma_l(\langle m|\mathbb{J}|x\rangle)\,. \tag{4.10}$$

Then how can we treat, for example, time-series data

$$|x(t)\rangle, \quad t = 1, 2, 3, \ldots, T, \tag{4.11}$$

within neural network models? Putting this one by one into the neural network is one way. However, as an example of (4.11), what if we have the following?

$$\begin{aligned}
|x(t=1)\rangle &= |\texttt{I}\rangle,\\
|x(t=2)\rangle &= |\texttt{have}\rangle,\\
|x(t=3)\rangle &= |\texttt{a}\rangle,\\
|x(t=4)\rangle &= |\texttt{pen}\rangle,\\
|x(t=5)\rangle &= |\texttt{.}\rangle.
\end{aligned} \tag{4.12}$$

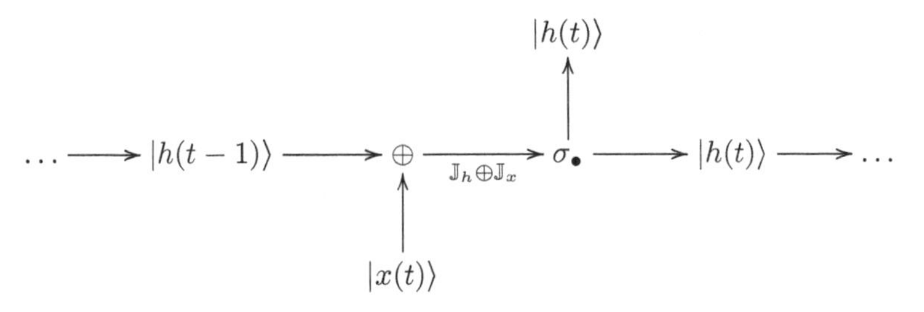

Fig. 4.7 Schematic diagram of a simple recurrent neural network

A model described so far that does not include the information of the time stamp before and after a certain item of data cannot capture the context. So as a simple extension let us consider

$$|h(t)\rangle = \sum_m |m\rangle \sigma_\bullet \Big(\langle m|\mathbb{J}_x|x(t)\rangle + \langle m|\mathbb{J}_h|h(t-1)\rangle \Big) . \tag{4.13}$$

This is the simplest form of what is called a recurrent neural network. See Fig. 4.7. Here, $|h(t)\rangle$ is the output value at each time t and the second input at the next time $t+1$. The first input is $|x(t+1)\rangle$. If we consider robot arm to carry luggage, $|x(t)\rangle$ is the image data coming in from the robot's field of view at each time, and $|h(t)\rangle$ is the movement of the arm at that time. In response to the arm movement, for example, if the arm is accelerated too much at time t, it is necessary to apply a brake at the next time step so that luggage will not be thrown. In this way, the recursive structure is included in order to deal with the case where we need to know what the previous behavior was in order to operate at the current time.

By the way, by adjusting $\mathbb{J}_h, \mathbb{J}_x, \sigma_\bullet$ well, even such a simple recurrent neural network can have arbitrary computational power. It is known that [43]. This is regarded as a kind of universal approximation theorem of neural networks, but in other words, it means that "recurrent neural networks can be used as a programming language," and that "it has the same capabilities as computers." Such a property is called **Turing-complete** or **computationally universal**. If a "general-purpose AI" is constructed (although it would be in the distant future as we write in 2019 January), it should have the ability to implement (acquire) any logical operation (language function) from data (experience) by itself. The computational completeness of recurrent neural networks reminds us somewhat of its possibility.

Backpropagation of error

Let us take a look at **backpropagation of error** in a recurrent neural network. We use the following recursive expression repeatedly:

$$
\begin{aligned}
&\delta|h(t)\rangle \\
&= \underbrace{\sum_m |m\rangle\sigma'_\bullet\Big(\langle m|\mathbb{J}_x|x(t)\rangle + \langle m|\mathbb{J}_h|h(t-1)\rangle\Big)\langle m|}_{\text{We name it } =:\mathbb{G}(t)} \begin{pmatrix} \delta\mathbb{J}_x|x(t)\rangle \\ +\delta\mathbb{J}_h|h(t-1)\rangle \\ +\mathbb{J}_h\delta|h(t-1)\rangle \end{pmatrix} \\
&= \mathbb{G}(t)\delta\mathbb{J}_x|x(t)\rangle + \mathbb{G}(t)\delta\mathbb{J}_h|h(t-1)\rangle + \mathbb{G}(t)\mathbb{J}_h\delta|h(t-1)\rangle\,.
\end{aligned}
\tag{4.14}
$$

Error function L is typically

$$
L = \langle d(1)|h(1)\rangle + \langle d(2)|h(2)\rangle + \dots\,, \tag{4.15}
$$

and it is just a summation of individual terms. So it would be sufficient to consider the tth term:

$$
\begin{aligned}
\delta\langle d(t)|h(t)\rangle &= \langle d(t)|\delta|h(t)\rangle \\
&= \underbrace{\langle d(t)|\mathbb{G}(t)}_{=:\langle\delta_t(t)|}\Big(\delta\mathbb{J}_x|x(t)\rangle + \delta\mathbb{J}_h|h(t-1)\rangle + \mathbb{J}_h\delta|h(t-1)\rangle\Big) \\
&= \langle\delta_t(t)|\delta\mathbb{J}_x|x(t)\rangle + \langle\delta_t(t)|\delta\mathbb{J}_h|h(t-1)\rangle \\
&\quad + \underbrace{\langle\delta_t(t)|\mathbb{J}_h\mathbb{G}(t-1)}_{=:\langle\delta_t(t-1)|}\Big(\delta\mathbb{J}_x|x(t-1)\rangle + \delta\mathbb{J}_h|h(t-2)\rangle + \mathbb{J}_h\delta|h(t-2)\rangle\Big) \\
&= \dots \\
&= \langle\delta_t(t)|\delta\mathbb{J}_x|x(t)\rangle + \langle\delta_t(t)|\delta\mathbb{J}_h|h(t-1)\rangle \\
&\quad + \langle\delta_t(t-1)|\delta\mathbb{J}_x|x(t-1)\rangle + \langle\delta_t(t-1)|\delta\mathbb{J}_h|h(t-2)\rangle \\
&\quad \dots \\
&\quad + \langle\delta_t(1)|\delta\mathbb{J}_x|x(1)\rangle + \langle\delta_t(1)|\delta\mathbb{J}_h|h(0)\rangle.
\end{aligned}
\tag{4.16}
$$

Here, using

$$
\delta\mathbb{J} = \sum_{m,n}|m\rangle\langle n|\delta J^{mn}\,, \tag{4.17}
$$

we find

$$\begin{aligned}\delta\langle d(t)|h(t)\rangle &= \sum_{\tau\leq t}\Big(\langle\delta_t(\tau)|\delta\mathbb{J}_x|x(\tau)\rangle + \langle\delta_t(\tau)|\delta\mathbb{J}_h|h(\tau-1)\rangle\Big)\\ &= \sum_{\tau\leq t}\sum_{m,n}\Big(\langle\delta_t(\tau)|m\rangle\langle n|x(\tau)\rangle\delta J_x^{mn} + \langle\delta_t(\tau)|m\rangle\langle n|h(\tau-1)\rangle\delta J_h^{mn}\Big)\\ &= \sum_{m,n}\Big(\sum_{\tau\leq t}\langle\delta_t(\tau)|m\rangle\langle n|x(\tau)\rangle\delta J_x^{mn} + \sum_{\tau\leq t}\langle\delta_t(\tau)|m\rangle\langle n|h(\tau-1)\rangle\delta J_h^{mn}\Big).\end{aligned} \tag{4.18}$$

Therefore,

$$\begin{aligned}\delta L &= \sum_t \delta\langle d(t)|h(t)\rangle\\ &= \sum_{m,n}\Big(\sum_t\sum_{\tau\leq t}\langle\delta_t(\tau)|m\rangle\langle n|x(\tau)\rangle\delta J_x^{mn} + \sum_t\sum_{\tau\leq t}\langle\delta_t(\tau)|m\rangle\langle n|h(\tau-1)\rangle\delta J_h^{mn}\Big),\end{aligned} \tag{4.19}$$

and then, the update rule is

$$\delta J_x^{mn} = -\epsilon\sum_t\sum_{\tau\leq t}\langle\delta_t(\tau)|m\rangle\langle n|x(\tau)\rangle, \quad \delta J_h^{mn} = -\epsilon\sum_t\sum_{\tau\leq t}\langle\delta_t(\tau)|m\rangle\langle n|h(\tau-1)\rangle. \tag{4.20}$$

Also,

$$\langle\delta_t(\tau-1)| = \langle\delta_t(\tau)|\mathbb{J}_h\mathbb{G}(\tau-1), \quad \langle\delta_t(t)| = \langle d(t)|\mathbb{G}(t) \tag{4.21}$$

is the formula for backpropagation in the present case.

Exploding gradient/vanishing gradient
Using the method explained above, we can train a recurrent neural network with data, but is that enough? For example, can we train the network using language data to make a good sentence? The answer is no. For example, it cannot handle parenthesis structure well. That is, loss of memory occurs. In the following, we explain why this happens in terms of the backpropagation equation.

Since we adjust the parameter J according to (4.20), let us start from those equations. $\langle\delta_t(\tau)|$ commonly appears in both update expressions. Considering the $\tau = 0$ state, from the backpropagation formula (4.21) we find

$$\begin{aligned}\langle\delta_t(1)| &= \langle\delta_t(2)|\mathbb{J}_h\mathbb{G}(1) = \langle\delta_t(3)|\mathbb{J}_h\mathbb{G}(2)\mathbb{J}_h\mathbb{G}(1) = \dots\\ &= \langle d(t)|\mathbb{G}(t)\underline{\mathbb{J}_h}\mathbb{G}(t-1)\dots\underline{\mathbb{J}_h}\mathbb{G}(2)\underline{\mathbb{J}_h}\mathbb{G}(1).\end{aligned} \tag{4.22}$$

Note that exactly the same $\mathbb{J}_h$ acts $t-1$ times from the right. Since t is the length of a sentence, it can be very long. Then

$$|\mathbb{J}_h| > 1\colon \langle \delta_t(1)| \text{ is too large,} \tag{4.23}$$

$$|\mathbb{J}_h| < 1\colon \langle \delta_t(1)| \text{ is too small.} \tag{4.24}$$

So both cases hinder learning. For example, if it is too large, it conflicts with the "small" updates, which is an implicit assumption that the update expression in (4.20) is valid. If it is too small, it means that $\tau = 0$ has no effect in the update formula, which means that the memory is forgotten. These are called **exploding gradient** and **vanishing gradient**, respectively. This was pointed out in Hochreiter's doctoral dissertation [44].[7]

4.3 LSTM

As explained above, with a simple recurrent neural network, the gradient explosion/vanishing cannot be improved by using the structure, deepening, convolution, etc. A new idea is needed to solve this problem. This section describes **LSTM** (long short-term memory) [46] which is commonly used currently.[8]

Memory vector

The core idea of LSTM is to set up a vector that controls the memory inside the recurrent neural network. It acts as a RAM (random access memory, temporary storage area). Let us call the memory vector at time t as

$$|c(t)\rangle \tag{4.25}$$

Fig. 4.8 shows the overall diagram of the LSTM.
Here g_f, g_i, and g_o are called

g_f: Forget gate

g_i: Input gate

g_o: Output gate

and we consider them as the sigmoid function for each component,

$$g_f = g_i = g_o = \sigma\,. \tag{4.26}$$

[7] This is written in German; for English literature see [45].

[8] An explanation without the bracket notation is found in [47] on which we largely rely.

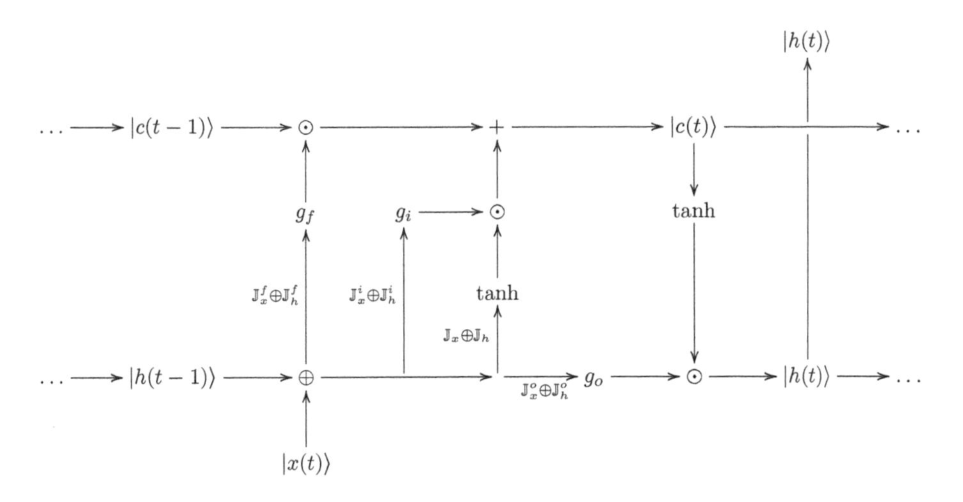

Fig. 4.8 Schematic diagram of LSTM. $\oplus$ represents the direct sum of two vectors in the input layer and $+$ represents the vector addition in the $|c(t)\rangle$ layer, and $\odot$ represents the multiplication for each component of the vector (and resultantly, the output vector has the same dimension)

The relations depicted in the diagram are summarized as follows:

$$|h(t)\rangle = \sum_m |m\rangle\langle m|g_o\rangle\langle m|\tanh(c(t))\rangle \,, \tag{4.27}$$

$$\begin{aligned} |c(t)\rangle &= \sum_m |m\rangle\langle m|c(t-1)\rangle\langle m|g_f\rangle \\ &\quad + \sum_m |m\rangle \tanh\Big(\langle m|(\mathbb{J}_x|x(t)\rangle + \mathbb{J}_h|h(t-1)\rangle)\Big)\langle m|g_i\rangle \,, \end{aligned} \tag{4.28}$$

$$|g_f\rangle = \sum_m |m\rangle g_f\Big(\langle m|(\mathbb{J}_x^f|x(t)\rangle + \mathbb{J}_h^f|h(t-1)\rangle)\Big) \,, \tag{4.29}$$

$$|g_i\rangle = \sum_m |m\rangle g_i\Big(\langle m|(\mathbb{J}_x^i|x(t)\rangle + \mathbb{J}_h^i|h(t-1)\rangle)\Big) \,, \tag{4.30}$$

$$|g_o\rangle = \sum_m |m\rangle g_o\Big(\langle m|(\mathbb{J}_x^o|x(t)\rangle + \mathbb{J}_h^o|h(t-1)\rangle)\Big) \,. \tag{4.31}$$

Let us see how the error backpropagates:

$$\delta|h(t)\rangle = \sum_m |m\rangle\Big[\langle m|\underbrace{\delta|g_o\rangle}_{(A)}\langle m|\tanh(c(t))\rangle + \langle m|g_o\rangle\langle m|\underbrace{\delta|\tanh(c(t))\rangle}_{(B)}\Big] \tag{4.32}$$

First, the first half (A) is transformed as follows:

$$
\begin{aligned}
(A) = \delta|g_o\rangle &= \delta \sum_m |m\rangle g_o\Big(\langle m|(\mathbb{J}_x^o|x(t)\rangle + \mathbb{J}_h^o|h(t-1)\rangle)\Big) \\
&= \sum_m |m\rangle g_o'(\bullet)\Big[\langle m|\delta|(\mathbb{J}_x^o|x(t)\rangle + \mathbb{J}_h^o|h(t-1)\rangle)\Big] \\
&= \sum_m |m\rangle g_o'(\bullet)\Big[\langle m|(\delta\mathbb{J}_x^o|x(t)\rangle + \delta\mathbb{J}_h^o|h(t-1)\rangle) + \mathbb{J}_h^o\delta|h(t-1)\rangle\Big].
\end{aligned} \tag{4.33}
$$

The third item is a form in which the parameter $\mathbb{J}_h^o$ acts on $\delta|h\rangle$ one time step before, so if we go back more and more time, we have a lot of $\mathbb{J}_h^o$, causing the exploding/vanishing gradient problem. Next, the second half (B) is written as

$$
\begin{aligned}
(B) &= \delta \sum_m |m\rangle \tanh(\langle m|c(t)\rangle) \\
&= \sum_m |m\rangle \tanh'(\bullet)\Big[\langle m|\underbrace{\delta|c(t)\rangle}_{(C)}\Big],
\end{aligned} \tag{4.34}
$$

so the magnitude of the gradient is left to the behavior of (C).

$$
\begin{aligned}
(C) &= \delta \sum_m |m\rangle\langle m|c(t-1)\rangle\langle m|g_f\rangle \\
&\quad + \delta \sum_m |m\rangle \tanh\Big(\langle m|(\mathbb{J}_x|x(t)\rangle + \mathbb{J}_h|h(t-1)\rangle)\Big)\langle m|g_i\rangle \\
&= \sum_m |m\rangle\Big[\langle m|\delta|c(t-1)\rangle\langle m|g_f\rangle + \langle m|c(t-1)\rangle\langle m|\underbrace{\delta|g_f\rangle}_{(D)}\Big] \\
&\quad + \sum_m |m\rangle\Big[\tanh'(\bullet)\Big(\langle m|(\delta\mathbb{J}_x|x(t)\rangle + \delta\mathbb{J}_h|h(t-1)\rangle + \mathbb{J}_h\delta|h(t-1)\rangle)\Big)\langle m|g_i\rangle \\
&\qquad + \tanh(\bullet)\langle m|\underbrace{\delta|g_i\rangle}_{(F)}\Big].
\end{aligned} \tag{4.35}
$$

Since (D) and (F) are basically the same as g_o, $\mathbb{J}^f$ and $\mathbb{J}^i$ act many times, possibly resulting in the exploding/vanishing gradient problem. However, in the first term of (4.35), $|c(t-1)\rangle$ does not have $\mathbb{J}$. This part is the backpropagation to the memory vector one time step ago, and plays the role of "keeping the memory as much as possible"—since the learning parameter $\mathbb{J}$ is not acting, in principle, it can propagate as long as it can. This is actually clear from the definition: $|x(t)\rangle$, $|h(t)\rangle$ is always input to the activation function, whereas $|c(t)\rangle$ is just multiplied by some constant by g_f. This is the most important part of the LSTM, the point where the exploding/vanishing gradient problem is less likely to occur.

Note that since the weight from the forget gate and the input from the input gate are placed at every τ, the information recorded once is not completely retained. In the memory vector, the "forget" and "remember" operations controlled only by $\mathbb{J}$ are given in advance in an explicit form. For example, let us take the following memory vector:

$$|c\rangle = \begin{pmatrix} c_1 \\ c_2 \\ \vdots \\ c_\spadesuit \end{pmatrix} \tag{4.36}$$

and assume that c_1 = (the component that governs a subject). If the input is `I enjoy machine learning.`, and a successfully trained LSTM is a machine that "predicts the next word," when the first word `I` comes in, the memory should be able to work as

$$|c\rangle = \begin{pmatrix} 1 \\ 0 \\ \vdots \\ 0 \end{pmatrix} \tag{4.37}$$

and recognize the situation as "we are dealing with the subject part no".[9] At this stage, it may be followed by `and you ...`, so it is not possible to cancel the "subject" state. When the second word `enjoy` is input, since this is a verb, the subject will no longer follow in English grammar. In this case, the output of the forget gate is

$$|g_f\rangle = \begin{pmatrix} 0 \\ 1 \\ \vdots \\ 1 \end{pmatrix}, \tag{4.38}$$

and when $|c\rangle$ is multiplied as $\odot$ with this, it is an operation to make c_1 = (the component that controls the subject) zero. Further, if the second component of the memory vector = (the component that controls the verb), the input gate when the second word `enjoy` is input is, for example,

[9]Keep in mind that this is just for illustration. In actual situations everything is dealt with in real numbers, and there are many kinds of different subjects. Here, consider the setting for convenience of explanation.

$$|g_i\rangle = \begin{pmatrix} 0 \\ 1 \\ 0 \\ \vdots \\ 0 \end{pmatrix} . \tag{4.39}$$

As above result, the memory shifts from the "subject" state to the "verb" state,[10]

$$|c(\text{Next time step})\rangle = |c\rangle \odot |g_f\rangle + |g_i\rangle$$
$$= \begin{pmatrix} 1 \\ 0 \\ 0 \\ \vdots \\ 0 \end{pmatrix} \odot \begin{pmatrix} 0 \\ 1 \\ 1 \\ \vdots \\ 1 \end{pmatrix} + \begin{pmatrix} 0 \\ 1 \\ 0 \\ \vdots \\ 0 \end{pmatrix} = \begin{pmatrix} 0 \\ 1 \\ 0 \\ \vdots \\ 0 \end{pmatrix} . \tag{4.40}$$

Attention mechanism
The operation to rate the importance of the feature vector according to the input, like the forget gate, is called as **attention mechanism**. It is known that the accuracy can be dramatically increased in natural language translation if an external attention mechanism is added to the LSTM [48]. Furthermore, there is a paper [49] which claims that the recurrent structure of a neural network is not even necessary, but only an attention mechanism is necessary, and in fact, it has shown high performance in natural language processing (NLP). The attention mechanism has also been shown to dramatically increase the accuracy of image processing [50, 51], and has been playing a key role in modern deep learning. Interested readers are encouraged to read, for example, [52] and its references.

Column: Edge of Chaos and Emergence of Computability

In electronic devices such as personal computers and smartphones which are indispensable for life in today's world, "why" can we do what we want? Behind this question, there exists a profound world that goes beyond the mere "useful tools" and is as good as any law of physics.

---

[10]Originally, tanh acts on the input vector as in (4.28), but it is omitted here for simplicity of the explanation.

Sorting Algorithm

First, let us consider sorting the following sequence in ascending order:

$$[3, 1, 2]. \tag{4.41}$$

Of course, the answer is [1,2,3], but if the sequence were much longer, like [38, 90, 25, 74, 87, 26, 53, 86, 14, 89, ...], it would be difficult to immediately answer. When we have to do this kind of work, one of the smartest ways is to write a program that does the sorting. In other words:

$$[3, 1, 2] \to \text{(Some "mechanical" procedure)} \to [1, 2, 3]. \tag{4.42}$$

The key term is "mechanical," in that there should be no human thinking during the procedure.[11] One way to implement this procedure is to pick up the input data two by two in order of precedence in the sequence, and swap if the right number is smaller, otherwise keep it as is:

$$[\underbrace{3,\ 1}_{1,3}, 2] \to [1, \underbrace{3,\ 2}_{2,3}] \to [1, 2, 3]. \tag{4.43}$$

It may not be complete just in a single trial. In another example, we find

$$[\underbrace{3,\ 2}_{2,3}, 1] \to [2, \underbrace{3,\ 1}_{1,3}] \to [2, 1, 3], \tag{4.44}$$

which is incomplete, and in such a case, the same process is performed again from the left:

$$(4.44) = [\underbrace{2,\ 1}_{1,2}, 3] \to [1, \underbrace{2,\ 3}_{2,3}] \to [1, 2, 3]. \tag{4.45}$$

Even if we do not want to judge whether or not "it has completed," the procedure should complete the sorting by repeating the same operation at least a number of times that is same as the number of elements of the sequence.

---

[11] A human is allowed to turn the gear by hand to put some energy to the machine.

Implementation Using Recurrent Neural Network

In the main text, we described that a **recurrent neural network** is Turing-complete. This means that if we adjust the training parameters of the network well, we can achieve something like (4.42). The paper [43] describes how to determine the weight by hand to achieve the desired processing, while a recent deep learning model (using a neural Turing machine (NTM) [53]) makes it possible to decide this automatically by learning. In other words, the network is becoming able to implement the process described above from the data itself.[12] The following shows the actual output of the trained network when an input that is not in the training data is input. Although it did not get it completely right because the training was not very good, it is still amazing to see the output coming out in the correct order at the correct length (though the values are slightly different).

$$[1, 1, 1, 2, 12] \rightarrow \begin{cases} \text{Correct answer } [1, 1, 1, 2, 12] \\ NTM \quad [1, 1, 1, 1, 12] \end{cases}$$

$$[13, 3, 1, 2, 5, 9, 9] \rightarrow \begin{cases} \text{Correct answer } [1, 2, 3, 5, 9, 9, 13] \\ NTM \quad [0, 3, 3, 3, 9, 13, 13] \end{cases}$$

$$[8, 10, 8, 3, 14, 4, 15, 5] \rightarrow \begin{cases} \text{Correct answer } [3, 4, 5, 8, 8, 10, 14, 15] \\ NTM \quad [0, 4, 6, 10, 10, 14, 14, 14] \end{cases}$$

$$[5, 2, 8, 12, 0] \rightarrow \begin{cases} \text{Correct answer } [0, 2, 5, 8, 12] \\ NTM \quad [2, 6, 6, 8, 12] \end{cases}$$

$$[5, 7, 12, 10, 2] \rightarrow \begin{cases} \text{Correct answer } [2, 5, 7, 10, 12] \\ NTM \quad [2, 6, 6, 8, 12] \end{cases}$$

KdV Equation and Box-Ball System

Let us introduce another story, a physical system called a box-ball system. The box-ball system is obtained by a specific discretization of the KdV equation[13] which is obtained from the fluid Navier-Stokes equation when the waves are

- shallow waves, and
- propagating in only one direction.

---

[12]The data is the supervised training data before and after the sorting. Here, about 800 data with a sequence length of up to 4 are randomly generated in binary notation, and an LSTM is used as a controller model.

[13]The name comes from Korteweg and de Vries.

Originally the system has spatially 1 dimension + time evolution, and now both are discretized. We draw the discretized space as

$$\cdots \circ\circ\bullet\bullet\circ\circ\circ\circ\circ\circ \cdots \tag{4.46}$$

In addition, ○ corresponds to the state where the wave is not standing, and ● corresponds to the state where the wave is standing. The unit time evolution of the box-ball system is

1. Select the leftmost ●
2. Make it ○ (and collect ●)
3. Move to the right by one unit
4. { If the state is ○ , put ■ there and return to 1
If the state is ● or ■, return to 3
5. Replace all ■ with ● (4.47)

The result of applying this unit time evolution to an initial configuration is as follows:

●●●○○○○○○○●●○○○○○○○●○○○○○○○○○○○○○○○○○○○○○○○○○○○○○○○

○○○●●●○○○○○○●●○○○○○○●○○○○○○○○○○○○○○○○○○○○○○○○○○○○○○

○○○○○○●●●○○○○○●●○○○○○●○○○○○○○○○○○○○○○○○○○○○○○○○○○○○

○○○○○○○○○●●●○○○○●●○○○○●○○○○○○○○○○○○○○○○○○○○○○○○○○○○

○○○○○○○○○○○○●●●○○○●●○○○●○○○○○○○○○○○○○○○○○○○○○○○○○○○

○○○○○○○○○○○○○○○●●●○○●●○○●○○○○○○○○○○○○○○○○○○○○○○○○○○

○○○○○○○○○○○○○○○○○○●●○○●●○●●○○○○○○○○○○○○○○○○○○○○○○○○

○○○○○○○○○○○○○○○○○○○○●●○○●○○●●●○○○○○○○○○○○○○○○○○○○○○

○○○○○○○○○○○○○○○○○○○○○○●●○●○○○○●●●○○○○○○○○○○○○○○○○○○

○○○○○○○○○○○○○○○○○○○○○○○○●○●●○○○○○●●●○○○○○○○○○○○○○○○

○○○○○○○○○○○○○○○○○○○○○○○○○●○○●●○○○○○○●●●○○○○○○○○○○○○

○○○○○○○○○○○○○○○○○○○○○○○○○○●○○○●●○○○○○○○●●●○○○○○○○○○

○○○○○○○○○○○○○○○○○○○○○○○○○○○●○○○○●●○○○○○○○○●●●○○○○○○

○○○○○○○○○○○○○○○○○○○○○○○○○○○○●○○○○○●●○○○○○○○○○●●●○○○

○○○○○○○○○○○○○○○○○○○○○○○○○○○○○●○○○○○○●●○○○○○○○○○○●●●

The readers may find that the time evolution similar to the scattering of solitons[14] of the KdV equations is obtained. By the way, if we count the number of • separated by ∘ from the left, this is $[3, 2, 1] \to [1, 2, 3]$, which is the same as the sorting algorithm! Just in case, if we put in another initial state corresponding to $[3, 4, 2, 1, 2, 3]$, then $[1, 2, 2, 3, 3, 4]$ is obtained:[15]

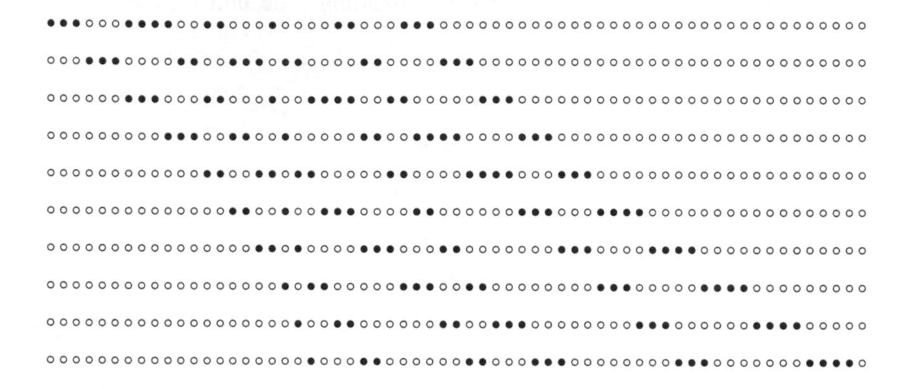

Namely, the physical phenomena described by the KdV equation seem to have included the sorting algorithm. This discrete-time evolution model of discrete states in a discrete space such as this box-ball system is called a cellular automaton.

Critical State and Turing Completeness of One-Dimensional Cellular Automata

Stephen Wolfram, famous for the mathematical software Mathematica, was originally a physicist working on elementary particle theories. He is also famous for classification of the one-dimensional (elementary) cellular automaton [54]. According to the classification, the dynamics of cellular automata are either:

I. Static (any initial state immediately stops and stabilizes)

II. Periodic (any initial state immediately stops or periodically moves and stabilizes)

III. Chaotic (the state does not stabilize even after sufficient time)

IV. Other than the above (the "edge" of stability and chaos)

[14]Solutions of differential equations with a localized energy density are called solitons (in physics).

[15]Note that if the solitons are not well-separated, calculation errors due to some phase shift will occur.

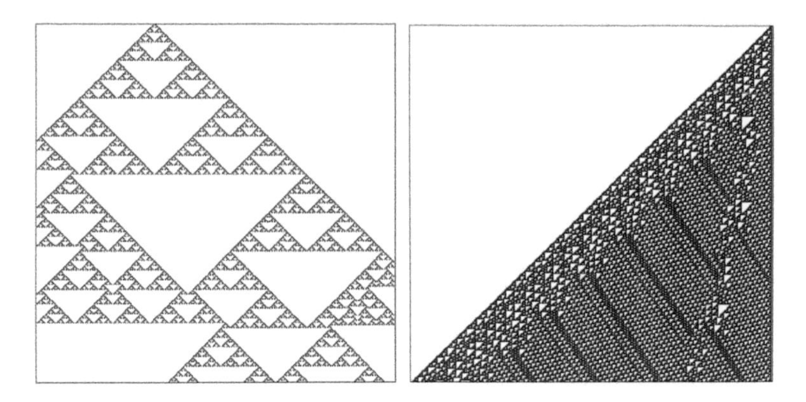

Fig. 4.9 Left: Rule 90, Right: Rule 110

For example, in the world of his "Rule 90," if only one point is set to •, an evolution diagram like Fig. 4.9 (left) can be obtained. This is a fractal figure called the famous Sierpiński triangle, and belongs to class III. On the other hand, "Rule 110" belongs to class IV, and it evolves over time as shown in Fig. 4.9 (right). We can see that various triangles of various sizes are generated, and in fact, it has been proved that this world is Turing-complete [55]. In other words, in this one-dimensional world, computers can be made. The more detailed explanation by the author of [55] is available in [56]. Christopher Langton, who is famous for cellular automata called Langton's ants, further argues that class IV time evolution corresponds to Turing-completeness due to its similarity to phase transition phenomena in condensed matter systems [57]. By the way, in the world we live in there are computers and they can perform calculation. Does that have anything to do with the physical laws of the universe? Research may advance, and we may come to answer such a question in the future.

Chapter 5
Sampling

Abstract In the situation where the training is performed, it is assumed that the input data is given by a probability distribution. It is often necessary to calculate the expectation value of the function of various input values given by the probability distribution. In this chapter, we will look at the method and the necessity of "sampling," which is the method of performing the calculation of the expectation value. The frequently used concepts in statistical mechanics, such as the law of large numbers, the central limit theorem, the Markov chain Monte Carlo method, the principle of detailed balance, the Metropolis method, and the heat bath method, are also used in machine learning. Familiarity with common concepts in physics and machine learning can lead to an understanding of both.

As we explained in Chap. 2, in this book we define machine learning as:

1. Assuming that there exists a probability distribution $P(x, d)$ that generates data
2. Adjust parameter J of probability distribution $Q_J(x, d)$ to approach $P(x, d)$

In order to introduce neural networks in Chaps. 3 and 4, we applied our knowledge of statistical mechanics to this point of view. The output of the deep neural network can be considered as the expectation value of d for the conditional probability distribution $Q_J(d|x)$ given the input x. However, one may often want to perform a sampling according to $Q_J(d|x)$ as the probability of occurrence for d, instead of taking the expectation value. For example, let us consider

$$Q_J(d|x) = \text{Probability of classifying what is in the input image } x \text{ as } d.$$

Suppose

$$x = \text{Photo of grandfather laughing with his dog},$$

A. Tanaka et al., *Deep Learning and Physics*, Mathematical Physics Studies,
https://doi.org/10.1007/978-981-33-6108-9_5

then we may find

$$Q_J(\text{grandfather}|x) = \frac{1}{2}, \quad Q_J(\text{dog}|x) = \frac{1}{2}.$$

Of course, this is fine, but if we sell this machine to our customers, one of them would complain as,[1]

Should x be classified as a grandfather or a dog?!

In such a case, one irresponsible solution is to use a coin toss:

$$\begin{aligned} d &= \text{grandfather, if the coin shows the front} \\ d &= \text{dog, if the coin shows the back} \end{aligned} \tag{5.1}$$

In this way we can respond to that complaint. Of course, based on (5.1), if the same image is used to ask the question many times, it will be judged sometimes as "grandfather" and sometimes as "dog." It will be no problem because showing the image to humans will result in the same.

The same problem occurs when actually creating the training data $\{(x[i], d[i])\}_i$, even before using the model Q_J which was trained. That is, when many objects a, b are shown in a single data item $x[i]$, the problem is whether the correct answer should be $d[i] = a$, or $d[i] = b$. Again, this depends on the judgment of the person creating the data, and that information is included in the data generation probability $P(x, d)$.

Although we have been silent until now about this issue, from the above considerations, we can see that it is necessary to consider what it means to actually sample when there is a probability distribution P. So, in this chapter, we explain the basics about and around the sampling, by focusing on important items related to:

- sampling of training data (data collection)
- sampling from trained machines (data imitation)

5.1 Central Limit Theorem and Its Role in Machine Learning

Let us recall the example given in Chap. 1 for the introduction to machine learning. We have possible events $A_1, A_2, \dots, A_W$, and the probability of occurrence is $p_1, p_2, \dots, p_W$, but we do not know the specific value of the probability p_i. Instead,

[1]This is, of course, a joke to illustrate the need for sampling.

we only know how many times each event occurred,

$$\bullet \begin{cases} A_1 : \#_1 \text{ times}, A_2 : \#_2 \text{ times}, \ldots, A_W : \#_W \text{ times}, \\ \# = \sum_{i=1}^{W} \#_i \text{ times in total.} \end{cases} \tag{5.2}$$

In Chap. 1, we used, without any proof, the fact that the maximum likelihood estimation converges to the desired probability p_i with in the limit of the number of data $\# \to \infty$,

$$\frac{\#_i}{\#} \to p_i, \tag{5.3}$$

which seems intuitively correct. However, when asked why it is fine to use (5.3), how can we answer it properly?

Law of large numbers
Here, consider the following:

$$X_n^{(i)} = \begin{cases} 1 \text{ if event } A_i \text{ occurs in the } n\text{th trial}, \\ 0 \text{ if event } A_i \text{ does not occur in the } n\text{th trial}. \end{cases} \tag{5.4}$$

Then, we can write

$$\frac{\#_i}{\#} = \frac{1}{\#} \sum_{n=1}^{\#} X_n^{(i)}. \tag{5.5}$$

Let us consider the "expectation value" of this ratio, which is

$$\begin{aligned} \left\langle \frac{\#_i}{\#} \right\rangle_p &= \left\langle \frac{1}{\#} \sum_{n=1}^{\#} X_n^{(i)} \right\rangle_p \\ &= \frac{1}{\#} \sum_{n=1}^{\#} \left\langle X_n^{(i)} \right\rangle_p \\ &= \frac{1}{\#} \sum_{n=1}^{\#} \Big(p_i \cdot \underbrace{1}_{X_n^{(i)}} + (1 - p_i) \cdot \underbrace{0}_{X_n^{(i)}} \Big) \\ &= \frac{1}{\#} \sum_{n=1}^{\#} p_i = p_i , \end{aligned} \tag{5.6}$$

and is equal to the probability of occurrence of event A_i. This is quite reminiscent of (5.3), but not exactly equal to (5.3) itself. So, let us investigate the "difference" between the actual $\frac{\#_i}{\#}$ and its expectation value $\left\langle \frac{\#_i}{\#} \right\rangle_p$. Note that the actually observed $\frac{\#_i}{\#}$ fluctuates, hence it does not make much sense just taking the difference. One way to remedy this situation is to consider the probability of having $\left\langle \frac{\#_i}{\#} \right\rangle_p - \epsilon \leq \frac{\#_i}{\#} \leq \left\langle \frac{\#_i}{\#} \right\rangle_p + \epsilon$ for a proper real positive constant ϵ. *If* we can show

$$P\left(\left|\frac{\#_i}{\#} - \left\langle \frac{\#_i}{\#} \right\rangle_p\right| < \epsilon\right) \geq 1 - \frac{♠}{\#\text{Positive number}}, \tag{5.7}$$

then for any ϵ, with a sufficiently large # (the number of data), this probability approaches unity, which proves (5.3). This is not a ϵ-δ definition of limit, but "ϵ-#" definition of limit. To show this, first consider the expectation value of the square of the difference:

$$\begin{aligned}
\left\langle \left|\frac{\#_i}{\#} - \left\langle \frac{\#_i}{\#} \right\rangle_p\right|^2 \right\rangle_p &= \sum_{\text{All possible}} P\left(\frac{\#_i}{\#}\right)\left|\frac{\#_i}{\#} - \left\langle \frac{\#_i}{\#} \right\rangle_p\right|^2 \\
&\geq \sum_{\left|\frac{\#_i}{\#} - \left\langle \frac{\#_i}{\#} \right\rangle_p\right| \geq \epsilon} P\left(\frac{\#_i}{\#}\right)\left|\frac{\#_i}{\#} - \left\langle \frac{\#_i}{\#} \right\rangle_p\right|^2 \\
&\geq \epsilon^2 \sum_{\left|\frac{\#_i}{\#} - \left\langle \frac{\#_i}{\#} \right\rangle_p\right| \geq \epsilon} P\left(\frac{\#_i}{\#}\right) \\
&= \epsilon^2 P\left(\left|\frac{\#_i}{\#} - \left\langle \frac{\#_i}{\#} \right\rangle_p\right| \geq \epsilon\right).
\end{aligned} \tag{5.8}$$

The first "$\geq$" came from restricting the range of the sum, and for the second "$\geq$", we used the condition of the restricted sum $\left|\frac{\#_i}{\#} - \left\langle \frac{\#_i}{\#} \right\rangle_p\right| \geq \epsilon$. The last "=" is the definition itself for $P\left(\left|\frac{\#_i}{\#} - \left\langle \frac{\#_i}{\#} \right\rangle_p\right| \geq \epsilon\right)$. In the calculation result here, compared with the desired probability is $P\left(\left|\frac{\#_i}{\#} - \left\langle \frac{\#_i}{\#} \right\rangle_p\right| < \epsilon\right)$, the inequality in parentheses is reversed. However, the sum of them should be unity, so we find

$$\begin{aligned}
P\left(\left|\frac{\#_i}{\#} - \left\langle \frac{\#_i}{\#} \right\rangle_p\right| < \epsilon\right) &= 1 - P\left(\left|\frac{\#_i}{\#} - \left\langle \frac{\#_i}{\#} \right\rangle_p\right| \geq \epsilon\right) \\
&\geq 1 - \frac{1}{\epsilon^2}\left\langle \left|\frac{\#_i}{\#} - \left\langle \frac{\#_i}{\#} \right\rangle_p\right|^2 \right\rangle_p.
\end{aligned} \tag{5.9}$$

Next, we get the following[2] expression of the second term on the right-hand side:

$$\begin{aligned}\left\langle\left|\frac{\#_i}{\#}-\left\langle\frac{\#_i}{\#}\right\rangle_p\right|^2\right\rangle_p &= \left\langle\left(\frac{\#_i}{\#}\right)^2-2\left(\frac{\#_i}{\#}\right)\left\langle\frac{\#_i}{\#}\right\rangle_p+\left\langle\frac{\#_i}{\#}\right\rangle_p^2\right\rangle_p \\ &= \left\langle\left(\frac{\#_i}{\#}\right)^2\right\rangle_p-2\left\langle\left(\frac{\#_i}{\#}\right)\right\rangle_p\left\langle\frac{\#_i}{\#}\right\rangle_p+\left\langle\frac{\#_i}{\#}\right\rangle_p^2 \\ &= \left\langle\left(\frac{\#_i}{\#}\right)^2\right\rangle_p-\left\langle\frac{\#_i}{\#}\right\rangle_p^2 \\ &= \frac{1}{\#}\Big(p_i(1-p_i)\Big),\end{aligned} \tag{5.11}$$

summarizing all, we find

$$P\left(\left|\frac{\#_i}{\#}-p_i\right|<\epsilon\right)\geq 1-\frac{p_i(1-p_i)}{\epsilon^2\#}. \tag{5.12}$$

This is what we wanted to show.

Law of large numbers (general case)
So far, we have considered how close the empirical probability is to the true probability p_i. In general, it is known[3] that for an "observable" X, the average of # trials

$$X^\# = \frac{1}{\#}\sum_{n=1}^{\#} X_n \tag{5.13}$$

is subject to the following condition:

$$P\Big(|X^\#-\mu|<\epsilon\Big)\geq 1-\frac{\sigma^2}{\epsilon^2\#}, \tag{5.14}$$

[2]The calculations of the first term can be done in the same way as (5.6):

$$\begin{aligned}\left\langle\left(\frac{\#_i}{\#}\right)^2\right\rangle_p &= \left\langle\left(\frac{1}{\#}\sum_{n=1}^{\#}X_n^{(i)}\right)^2\right\rangle_p = \left\langle\frac{1}{\#^2}\sum_{n=1,m=1}^{\#}X_n^{(i)}X_m^{(i)}\right\rangle_p \\ &= \frac{1}{\#^2}\sum_{n=1,m=1}^{\#}\left\langle X_n^{(i)}X_m^{(i)}\right\rangle_p = \frac{1}{\#^2}\sum_{n=1,m=1}^{\#}\begin{cases}p_i & (m=n)\\ p_i^2 & (m\neq n)\end{cases} \\ &= \frac{1}{\#^2}\Big(\#p_i+\#(\#-1)p_i^2\Big) = \frac{1}{\#}\Big(p_i(1-p_i)\Big)+p_i^2.\end{aligned} \tag{5.10}$$

The point is that there is a case division in the calculation.

[3]The derivation in the general case is left to the reader.

with

$$\mu = \langle X \rangle, \quad \sigma^2 = \langle (X - \mu)^2 \rangle . \tag{5.15}$$

In other words, we can use a sampling average $X^\#$ instead of the expectation value $\mu = \langle X \rangle$, if the number of samples # is enough large. So, to say it without worrying about being misunderstood, there is no need to calculate the expectation value when a large number of samples is possible. In fact, it is customary to calculate the average value of observables by sampling instead of the exact expectation value in the statistical mechanics of multi-degree-of-freedom systems in particle physics and condensed matter physics. When the probability that $X^\#$ converges to a certain value μ at $\# \to \infty$ as in (5.14) is unity, we say that $X^\#$ converges in probability to μ.

Central limit theorem

Now let us bring back the customer in the previous example:

I know that (5.14) holds, but what's the value of $X^\#$ after all?

Such an opinion would be natural in a sense. Equation (5.14) states that the sampling average approaches the expectation value (that is a convergence in probability), but does not tell how it approaches to the value. The central limit theorem actually tells that to us:

$$P(X^\#) \to \mathcal{N}\left(\mu, \frac{\sigma^2}{\#}\right), \tag{5.16}$$

where $\mathcal{N}(\mu, \sigma^2)$ is a Gaussian distribution with mean μ and variance σ^2. In other words, when the number of samples # is large, the sample average $X^\#$ follows a Gaussian distribution whose mean is the desired expectation value and whose variance is the variance of the original observation divided by the number of samples #. This is called convergence in distribution.[4] To be more specific,[5]

$$X^\# \text{ is approximately 70\% likely in the interval } \left[\mu - \frac{\sigma}{\sqrt{\#}}, \mu + \frac{\sigma}{\sqrt{\#}}\right]. \tag{5.17}$$

[4]It is also called convergence in law.

[5]This period is called 1σ. In particle physics, an observation with confidence of about $3\sigma = 99.73\%$ is called "evidence" and an observation with confidence of about $5\sigma = 99.99994\%$ is called "discovery" [58].

This can be derived by Fourier transforming the probability distribution,[6]

$$\begin{aligned}
\int dX^{\#} e^{itX^{\#}} P(X^{\#}) &= \langle e^{itX^{\#}} \rangle_{X^{\#}} \\
&= e^{it\mu} \langle \prod_{n=1}^{\#} e^{it\frac{1}{\#}(X_n - \mu)} \rangle_{X^{\#}} \\
&= e^{it\mu} \prod_{n=1}^{\#} \Big\langle \Big(1 + it\frac{1}{\#}(X_n - \mu) - \frac{t^2}{2\#^2}(X_n - \mu)^2 + \dots \Big) \Big\rangle_{X_n} \\
&= e^{it\mu} \prod_{n=1}^{\#} \Big(1 + \frac{1}{\#} \underbrace{\Big(\frac{-t^2}{2\#}\sigma^2 + \dots\Big)}_{\spadesuit} \Big) = e^{it\mu} \Big(1 + \frac{\spadesuit}{\#} \Big)^{\#} \\
&\to e^{it\mu + \spadesuit} .
\end{aligned} \tag{5.18}$$

Then the inverse transformation (which should return to the original probability distribution for $X^{\#}$) of this is

$$\begin{aligned}
&\frac{1}{2\pi} \int dt \, e^{-itX^{\#}} e^{it\mu + \spadesuit} \\
&= \frac{1}{2\pi} \int dt \, e^{-itX^{\#}} e^{it\mu - \frac{t^2}{2\#}\sigma^2 + \dots} \\
&= \frac{1}{2\pi} \int dt \exp\Big(-it(X^{\#} - \mu) - \frac{t^2}{2\#}\sigma^2 + \dots \Big) \\
&= \frac{1}{2\pi} \int dt \exp\Big(-\frac{\sigma^2}{2\#}(t + i\#\frac{X^{\#} - \mu}{\sigma^2})^2 - \frac{\#}{2\sigma^2}(X^{\#} - \mu)^2 + \dots \Big) \\
&\approx \sqrt{\frac{\#}{2\pi\sigma^2}} \exp\Big(-\frac{\#}{2\sigma^2}(X^{\#} - \mu)^2 \Big) .
\end{aligned} \tag{5.19}$$

This is indeed a Gaussian distribution with mean μ and variance $\frac{\sigma^2}{\#}$, $\mathcal{N}\Big(\mu, \frac{\sigma^2}{\#}\Big)$. Note that according to [59], the same conclusion can be drawn using the idea of a "renormalization group" in physics. We recommend interested readers to take a look at it.

[6] In statistics, it is called a characteristic function.

Meaning of central limit theorem in machine learning
Let us go back to the first example of the law of large numbers. When the true probability is not given while only (5.2) is given, the most likely estimation q_i of the true probability is

$$q_i = \frac{\#_i}{\#} \tag{5.20}$$

(see footnote 5 in Chap. 1). According to the central limit theorem, this "best" estimation q_i is subject to

$$\text{With approximately 70\% possibility, } p_i - \sqrt{\frac{p_i(1-p_i)}{\#}} < q_i < p_i + \sqrt{\frac{p_i(1-p_i)}{\#}}. \tag{5.21}$$

When regarding (5.20) as the result of machine learning, (5.21) is considered to represent the generalization performance of this "machine." In order to have the generalization, that is, to make q_i very close to p_i, it is better to take a larger sample number #. This is reminiscent of the generalized performance inequality (2.11) using the VC dimensions described in Chap. 2. Actually, the $\sqrt{\#}$ in the denominator of the second term on the right-hand side of the inequality (2.11) of generalization performance is the same as (5.21), and it suggests that the part comes from the central limit theorem. Thus, the central limit theorem is closely related to generalization performance.

5.2 Various Sampling Methods

Next, let us look at models. In this book, the model of supervised machine learning is a conditional probability with parameters,

$$Q_J(d|x)\,. \tag{5.22}$$

When we actually "operate" this model, we want to label some input according to its probability. Here, there exists a gap between knowing the value of the probability and sampling according to that probability. For example, if the reader wants to create a machine that outputs one of 1, 2, 3, 4, 5, 6 with a probability of 1/6, then what does she/he do? The easiest solution is to make a dice, and for the sampling, just roll the dice. However, in the model (5.22), the probability of d varies for each input x. It is ridiculous to keep making dice of various shapes each time.

To make matters worse, there are machine learning models that make it more difficult to calculate the value of the probability in the first place, as described in the next chapter. Even in such a situation, there are cases in which sampling is possible.[7] As long as sampling is possible, the expectation value can be calculated by a method that replaces the expectation value with the sample average value, based on the law of large numbers (5.14) and the central limit theorem (5.16).

Thus, sampling techniques are an important concept to understand in machine learning methods. In the following, assuming that random numbers (sampling from a uniform probability) are given,[8] we explain some methods for performing sampling from more complex probabilities.

5.2.1 *Inverse Transform Sampling*

Assuming that it is possible to sample z which follows a uniform distribution, how can we sample x following a probability distribution $P(x)$? First, the probability of finding x is

$$P(x)dx\,. \tag{5.23}$$

Suppose that this is written as a total derivative of certain function $F(x)$ as follows:

$$dF(x) = P(x)dx. \tag{5.24}$$

Since the probability of finding z following a uniform distribution is proportional to dz, (5.24) means that, with regarding $F(x) = z$ and using the inverse function F^{-1} of F,

$$x := F^{-1}(z)\,, \tag{5.25}$$

[7] Although not described in this book, the calculation of a probability distribution called a posterior distribution in Bayesian estimation corresponds to this case.

[8] As a matter of fact, it is impossible to make a true random number with a classical calculator. A definition of random numbers can be given by Kolmogorov complexity, so it is not possible to provide a random sequence of infinite length. In many cases in physics, a reproducible sequence of random numbers is required for practical use, and so, pseudo-random numbers, which are sequences of finite period numbers that can be reproduced and are statistically unbiased, are used. Von Neumann said: "Any one who considers arithmetical methods of producing random digits is, of course, in a state of sin." But it is used in modern times with a lot of ingenuity. It is also known that a statistically biased pseudo-random number generation method can lead to incorrect calculations of the transition temperature of the simulated Ising model [60]. Not only that it should be statistically unbiased, but also a long periodicity where the same sequence does not appear is also required for large-scale simulations. More recently, the one published by Makoto Matsumoto and Takuji Nishimura, called Mersenne Twister, has been used worldwide [61]. It has advantages such as long periods, uniform distribution, and fast generation. Other pseudo-random number generators, such as XorShift and MIXMAX, have also been proposed [62]. We suggest that readers refer to other references for more information [63].

where x is a sampling from the desired probability distribution $P(x)$. This is called the inverse transform sampling. To actually execute the inverse transform method, it is necessary that F and F^{-1} can be calculated analytically.

Box–Muller method
As a specific example of the inverse transform method, let us consider how we can achieve a sampling of (x, y) from a two-dimensional Gaussian distribution

$$P(x, y) = \frac{1}{2\pi} e^{-\frac{1}{2}(x^2+y^2)} . \tag{5.26}$$

It is recommended to remember the derivation of the Gaussian integral with the polar coordinates. First, consider the two-dimensional polar coordinates

$$x = r\cos\theta , \quad y = r\sin\theta . \tag{5.27}$$

Then, with $\lambda = \frac{r^2}{2}$, the probability density is

$$P(x, y)dxdy = \frac{d\theta}{2\pi} \cdot e^{-\lambda} d\lambda . \tag{5.28}$$

Hence, θ is found to follow a uniform distribution of $[0, 2\pi]$. On the other hand, λ now follows a probability distribution called the exponential distribution on the $[0, \infty]$ interval. Since it is assumed that a uniform distribution is possible, if sampling from λ is possible, then set $r = \sqrt{2\lambda}$, and apply (5.27), then (x, y) is sampled. The inverse transform method can be used to sample λ. In fact, since

$$e^{-\lambda} d\lambda = d(-e^{-\lambda}) , \tag{5.29}$$

we find $F(\lambda) = -e^{-\lambda}$. The inverse function is obtained by solving $z = F(\lambda) = -e^{-\lambda}$ for λ, as

$$\lambda = -\log(-z) = F^{-1}(z) . \tag{5.30}$$

So this value should be λ. This is called the Box–Muller method [64]. In this way, from the uniform distribution sampling, by using the inverse transform method, the sampling is possible in the case when we can find $F(x)$ satisfying (5.24), namely in the case of a probability density function whose indefinite integral can be exactly written down.

5.2.2 Rejection Sampling

Unfortunately, the indefinite integral of the probability density function $P(x)$ is not always known. Even in unknown case, we can use the rejection sampling method. In the rejection sampling, we assume that a sampling from some probability distribution $Q(x)$ is possible. For example, we may use $Q(x)$ for that we can use the inverse transform method. Another assumption to use this method is

$$\text{There exists some number } M(>0), \text{ and for any } x,\ MQ(x) \geq P(x). \tag{5.31}$$

As long as this is satisfied, the sampling from $P(x)$ is possible, as follows:

1. $x_{\text{candidate}}$ is sampled according to $Q(x)$.
2. If $\frac{P(x_{\text{candidate}})}{MQ(x_{\text{candidate}})}$ is larger than a random number $0 < r < 1$, the sample $x_{\text{candidate}}$ is adopted. This is called *accepted*. If the above condition is not met, we discard $x_{\text{candidate}}$. This is called *rejected*.[9]

To understand why this method works, we use Bayes' theorem (see the column in Chap. 2) to find the x probability distribution for the acceptance,

$$P(x|\text{accepted}) = \frac{P(\text{accepted}|x)Q(x)}{P(\text{accepted})}. \tag{5.32}$$

This is because

$$P(\text{accepted}|x) = \frac{P(x)}{MQ(x)} \tag{5.33}$$

and we substitute

$$P(\text{accepted}) = \int P(\text{accepted}|x)Q(x)dx = \frac{1}{M} \tag{5.34}$$

into (5.32) to find

$$P(x|\text{accepted}) = P(x). \tag{5.35}$$

Collecting only the accepted $x_{\text{candidate}}$ results in a sampling from the probability distribution that we want. With this method, now we can sample from a fairly large class of probability distributions.

[9]Note that this operation is different from the rejection by the Metropolis method that will be explained later.

Weakness of the rejection sampling

Equation (5.31) is required for rejection sampling to work. Also, even if (5.31) is satisfied, if the value of M is very large, the possibility of rejection increases, and sampling cannot be performed. Furthermore, in the rejection sampling, the calculated $x_{\text{candidate}}$ may be simply discarded, so if it takes time for one sampling from $Q(x)$, it is not very efficient to collect samples. In order for the rejection sampling to be practical, we need to find a good $Q(x)$ for the desired $P(x)$.

For example, consider $x = s_i$ as a spin configuration, and let $P(x) = P(s_i)$ be the **Ising model**[10] in statistical mechanics. The Ising model is a classical model of a magnet, in which each point labeled i on the lattice has a variable (spin) that can take the value of 1 or -1. In the case of a two-dimensional Ising model whose lattice size is $L_x \times L_y$, the label i runs from 1 to $2^{L_x L_y}$. Specifically, for example, even with $L_x = L_y = 10$, we have $2^{10\times 10} \approx 1.27 \times 10^{30}$, and so we have to consider the probability distribution in a huge region. Also, the higher the dimensions, the more the label i can run, in the exponential form.[11] This phenomenon is called **the curse of dimensionality**. In such a case, it is very difficult to come up with a good proposal distribution $Q(s_i)$.

5.2.3 *Markov Chain*

A useful sampling technique even in such cases is to use **Markov chains**. In this subsection, we will explain the basics of Markov chain using concrete examples [65]. In particular, we will introduce the Markov chain with two examples, the **Gothenburg weather model** and the **Los Angeles weather model**. These examples here do not have the curse of dimensionality as described above, but do not lose the essence of the argument.

Markov chain

From now on, we will introduce the Markov chain through concrete examples. In particular, let us focus on the day's weather at a certain location, and for simplicity, assume that the weather is only rainy or sunny. The weather here is what is called a "state." According to the literature, we take $s_1 =$ rainy, $s_2 =$ sunny. In addition, let us introduce a vector $\mathbf{P}$ that represents the probability distribution for the two states. If the probability of rainy weather is P_{rainy} and the probability of sunny weather is P_{sunny}, we align them as

$$\mathbf{P} = \begin{pmatrix} P_{\text{rainy}} \\ P_{\text{sunny}} \end{pmatrix}. \tag{5.36}$$

[10] Readers unfamiliar with the Ising model should refer to the column.

[11] If L is the length of one side of the d-dimensional square lattice, the range of i is 2^{L^d}. For example, if one side of the 3-dimensional cube has 10 lattice points, the range of i is $2^{10^3} \approx 1.07 \times 10^{301}$.

This is a probability distribution, so note the relation $\sum_i P_i = 1$.

The weather at a certain location changes daily, as $s_1 \to s_2 \to s_1 \to s_1 \to \cdots$. Consider a simple model to predict tomorrow's weather. For example, if we make the assumption that the day after a sunny day will be more sunny and the day after a rainy day will be more rainy, this is a model called a Gothenburg weather model. This is a Markov chain, as described below.

Markov chains are defined by transitions between states. However, we consider only those whose state transitions are only dependent on the current state and not on the past history. The above weather model is a Markov chain, assuming that the next day's weather depends only on the previous day's weather.

At this time, we can introduce what is called a **transition matrix** that gives the "probability distribution of the next day's weather" from the "probability distribution of the previous day":

$$\mathbb{T}_{\mathrm{G}} = \begin{pmatrix} 0.75 & 0.25 \\ 0.25 & 0.75 \end{pmatrix}. \tag{5.37}$$

Here, the probability that the next day is the same weather is 75%, and the probability that the next day is different weather is 25%.[12]

Next, let us see how to use the transition matrix. Give an initial state, we will explain how it transforms, using an example. We start with the rainy state as an initial state:

$$P_0 = \begin{pmatrix} 1 \\ 0 \end{pmatrix}. \tag{5.38}$$

Then the next day's weather probability distribution $\mathbf{P}_1$ can be calculated using the transition matrix as follows:

$$\mathbf{P}_1 = \mathbb{T}_{\mathrm{G}} P_0 = \begin{pmatrix} 0.75 \\ 0.25 \end{pmatrix}. \tag{5.39}$$

In other words, the probability distribution of the next day's weather shows that the probability of rainy is 75% and the probability of sunny is 25%. Considering the next day's weather probability distribution $\mathbf{P}_2$,

$$\mathbf{P}_2 = \mathbb{T}_{\mathrm{G}}^2 P_0 = \begin{pmatrix} 0.625 \\ 0.375 \end{pmatrix}. \tag{5.40}$$

[12]In order to satisfy the probability normalization, summation for column must be 1.

It can be expected that the probability of having a sunny day is about 40%. If the probability distribution after n days is $\mathbf{P}_n$,

$$\mathbf{P}_n = \lim_{n\to\infty} P_0 = \begin{pmatrix} \frac{1+2^{-n}}{2} \\ \frac{1-2^{-n}}{2} \end{pmatrix}, \tag{5.41}$$

and in the limit $n \to \infty$,

$$\mathbf{P}_\infty = \begin{pmatrix} \frac{1}{2} \\ \frac{1}{2} \end{pmatrix}. \tag{5.42}$$

We conclude that we do not know if the weather is rainy or sunny in the distant future.

By the way, the maximum eigenvalue of the transition matrix $\mathbb{T}_\mathrm{G}$ is 1, and the eigenvector corresponding to the maximum eigenvalue is

$$\mathbf{v}_\mathrm{G} = \begin{pmatrix} \frac{1}{2} \\ \frac{1}{2} \end{pmatrix}, \tag{5.43}$$

which is identical with $\mathbf{P}_\infty$.[13] It is not coincidence that $\mathbf{P}_\infty$ matches $\mathbf{v}_\mathrm{G}$, and it is actually a special case of a theorem called the **Perron–Frobenius theorem**.

Let us explain the consequences of the Perron–Frobenius theorem for a matrix A whose element is non-negative.[14]

Perron-Frobenius theorem:

Let A be a matrix whose elements are positive real numbers, and irreducible[15] and suppose the eigenvalues are non-negative. Also assume that for a positive integer n, each component of A^n is greater than zero.[16] Then A has a positive eigenvalue α (Perron-Frobenius root) that satisfies:

1. For eigenvalues λ other than the eigenvalue α for A, $|\lambda| < \alpha$.
2. α is not degenerate, and all components of its eigenvector are positive.

[13] The other eigenvector $\mathbf{v}'_\mathrm{G}$ is

$$\mathbf{v}'_\mathrm{G} \propto \begin{pmatrix} 1 \\ -1 \end{pmatrix}, \tag{5.44}$$

which does not satisfy the positive semidefiniteness of the probability.

[14] Note the difference between the matrix A being positive definite and each element being positive. A matrix A is positive definite if its eigenvalues are all positive when diagonalized, and is defined as a property that is invariant under similarity transformation. On the other hand, the property that each element is positive is not invariant under the similarity transformation.

[15] This means that the matrix is not block-diagonal.

[16] This is the assumption of aperiodicity in Markov chains.

In the previous example, the probability distribution of $n \to \infty$ and the eigenvector of $\mathbb{T}_G$ matched because of the first property. When multiplying $\mathbb{T}_G$ many times, the non-degenerate eigenvector belonging to the largest eigenvalue is extracted.[17] The components of the eigenvector are positive and can be interpreted as a probability distribution, all of which are guaranteed by the Perron-Frobenius theorem.

Next, consider a model called the Los Angeles weather model. This is an asymmetrical model, according to the idea that there will be many sunny days. The transition matrix is given as:

$$\mathbb{T}_{LA} = \begin{pmatrix} 0.5 & 0.1 \\ 0.5 & 0.9 \end{pmatrix}. \tag{5.45}$$

Let the initial state be $P_0 = s_1$ again and consider the next day's weather probability distribution:

$$\mathbf{P}_1 = \mathbb{T}_{LA} P_0 = \begin{pmatrix} 0.5 \\ 0.5 \end{pmatrix}. \tag{5.46}$$

There is a probability distribution of a rainy day 50% and a sunny day 50%. For the next day,

$$\mathbf{P}_2 = \mathbb{T}_{LA}^2 P_0 = \begin{pmatrix} 0.3 \\ 0.7 \end{pmatrix}. \tag{5.47}$$

and it shows that the probability of a sunny day is high. Since each element of $\mathbb{T}_{LA}$ is positive, the Perron-Frobenius theorem can be applied. Therefore, the eigenvector of $\mathbb{T}_{LA}$ is calculated, and the probability distribution at $n \to \infty$ is calculated. The eigenvector $\mathbf{v}_{LA}$ belonging to the largest eigenvalue is

$$\mathbf{v}_{LA} = \begin{pmatrix} 1/6 \\ 5/6 \end{pmatrix} \approx \begin{pmatrix} 0.17 \\ 0.83 \end{pmatrix}. \tag{5.48}$$

That is,

$$\mathbf{P}_\infty \approx \begin{pmatrix} 0.17 \\ 0.83 \end{pmatrix}. \tag{5.49}$$

Therefore, the probability of being sunny in the distant future is about 83%.

[17]For those who are good at numerical calculations, remember the power method used for eigenvalue calculations.

5.2.4 *Master Equation and the Principle of Detailed Balance*

In this subsection, we derive the **principle of detailed balance**, a **sufficient condition** for a Markov chain to converge to a probability distribution. The previous subsection introduced a Markov chain for two states, and this subsection deals with N states. (States labeled by continuous variables are not described here.) For later purposes, we change the label of the number of steps from n to t.

Given a state, the probability of giving it should be defined, so we define the probability as a function of the state:

$$[\text{Probability that state } s_i \text{ is realized at step } t \text{ of Markov chain}] \equiv P(s_i; t). \tag{5.50}$$

Since we are at least in one of the states at each step t, we require the following:

$$1 = \sum_i P(s_i; t). \tag{5.51}$$

Let $P(s_j|s_i)$ be the transition probability from state s_i to s_j.[18] Let us require that all probabilities at the step t and the next step $t + \Delta t$ satisfy the following conservation law of probability. Recalling the conservation law of energy and that of charge in fluid mechanics and electromagnetism in a unit spatial volume, since the change of s_i = the amount flowing out of $-s_i$ + the amount coming in s_i, the conservation law should be

$$\frac{P(s_i; t + \Delta t) - P(s_i; t)}{\Delta t} = -\sum_{j \neq i} P(s_i; t) P(s_j|s_i) + \sum_{j \neq i} P(s_j; t) P(s_i|s_j). \tag{5.52}$$

By slightly modifying this, we get the probability of realizing the state s_i at the step $t + \Delta t$,

$$P(s_i; t + \Delta t) = P(s_i; t) - \sum_{j \neq i} P(s_i; t) P(s_j|s_i) \Delta t + \sum_{j \neq i} P(s_j; t) P(s_i|s_j) \Delta t. \tag{5.53}$$

The second term on the right-hand side is the probability of exiting the state s_i, and the third is the probability of becoming the state s_i.

Next, we introduce a probability vector for the N states,

$$\mathbf{P}(t) = \begin{pmatrix} P(s_1; t) \\ P(s_2; t) \\ \vdots \\ P(s_N; t) \end{pmatrix}. \tag{5.54}$$

[18] This is a conditional probability, but in this context it is called a transition probability.

Using the probability vector $\mathbf{P}(t)$, the change in the distribution (5.53) is written as

$$\mathbf{P}(t + \Delta t) = \mathbb{T}\mathbf{P}(t)\,. \tag{5.55}$$

Here the transition matrix $\mathbb{T}$ is

$$\mathbb{T} = \begin{pmatrix} a_{11} & \cdots & a_{1N} \\ \vdots & \ddots & \vdots \\ a_{N1} & \cdots & a_{NN} \end{pmatrix}, \tag{5.56}$$

in which the relation to the transition probability $P(s_j|s_i)$ is made as

$$a_{ij} = P(s_i|s_j)\Delta t \quad i \neq j\,, \tag{5.57}$$

$$a_{ii} = 1 - \sum_{j \neq i} P(s_j|s_i)\Delta t\,. \tag{5.58}$$

In the following, we will take $\Delta t = 1$.

Supposing that $\mathbf{P}(0)$ is a certain probability vector, we assume

$$\mathbb{T}^n \mathbf{P}(0) \to \mathbf{P}_{\text{eq}}\,, \tag{5.59}$$

in the limit of $n \to \infty$.[19] In other words, we assume that there exists a convergence destination for the probability vector and that it is reached. In this context, the probability distribution $\mathbf{P}_{\text{eq}}$ that appears in this limit is called the equilibrium distribution. If the equilibrium distribution $\mathbf{P}_{\text{eq}}$ appears in the above limit, then it satisfies

$$\mathbb{T}\mathbf{P}_{\text{eq}} = \mathbf{P}_{\text{eq}}\,. \tag{5.60}$$

This is written with the matrix elements as

$$\sum_j a_{ij} P_{\text{eq}}(s_j) = P_{\text{eq}}(s_i)\,, \tag{5.61}$$

Further dividing the sum on the left-hand side into j and i, it becomes

$$\sum_{j \neq i} a_{ij} P_{\text{eq}}(s_j) + a_{ii} P_{\text{eq}}(s_i) = P_{\text{eq}}(s_i)\,. \tag{5.62}$$

[19] In general cases, in order for a Markov chain $\mathbb{T}$ whose elements are all non-negative to uniquely converge, all transitions must satisfy the following two conditions [66]:

1. Irreducible (ergodic). That is, $\mathbb{T}$ is not a block diagonal.
2. Non-periodicity. That is, there is no state that always appears at a constant period.

And substituting the definition of a_{ij}, we obtain

$$\sum_{j\neq i} P(s_i|s_j)P_{\text{eq}}(s_j) + (1 - \sum_{j\neq i} P(s_j|s_i))P_{\text{eq}}(s_i) = P_{\text{eq}}(s_i)\,. \tag{5.63}$$

Then the equation that gives the convergence condition is found as

$$\sum_{j\neq i} \Big(P(s_i|s_j)P_{\text{eq}}(s_j) - P(s_j|s_i)P_{\text{eq}}(s_i) \Big) = 0\,. \tag{5.64}$$

This is called the **master equation**.

One of the solutions of this master equation is the principle of detailed balance, as described below:

$$P(s_i|s_j)P_{\text{eq}}(s_j) = P(s_j|s_i)P_{\text{eq}}(s_i). \tag{5.65}$$

Note that this is a sufficient condition for the convergence of Markov chains.

The principle of detailed balance gives a sufficient condition for the convergence of Markov chains, and there is a freedom in choosing the transition probability $P(s_i|s_j)$ that appears there. In the following, we first explain the Markov chain Monte Carlo method, which is one of the applications, and in the next section we will introduce a typical choice of $P(s_i|s_j)$.

Here, let us mention the difference between the weather problem mentioned in the previous section and the Markov chain Monte Carlo method. In the weather model, given the transition matrix $\mathbb{T}$, the convergence destination P_{eq} was determined later. On the other hand, when using the Markov chain Monte Carlo method, P_{eq} is known, and $\mathbb{T}$ is designed accordingly. The guideline for the design is the principle of the detailed balance.

5.2.5 Expectation Value Calculation Using Markov chains, and Importance Sampling

From the above discussion, it was found that the equilibrium distribution P_{eq} can be obtained by using a Markov chain satisfying appropriate conditions. In the actual calculations, it is not possible to take the exact limit $n \to \infty$, so the calculation will be performed in the following procedures.

First, we consider the probability distribution after starting from an appropriate state and making enough transitions. In other words, for $k \gg 1$, we consider

$$\mathbf{P}_{\text{eq}} \approx \mathbb{T}^k \mathbf{P}_0 = \mathbf{P}^{(k)}_{\text{eq,app}}\,. \tag{5.66}$$

Then, from convergence and $k \gg 1$, we expect that $\mathbf{P}^{(k)}_{\text{eq,app}}$ is sufficiently close to $\mathbf{P}_{\text{eq}}$.[20] In other words, the state obtained at this point is considered as a sampling from the equilibrium states. In practice, it is difficult to judge, but if one calculates the expectation value at each step and observes that it starts to fluctuate around a certain central value, it is considered that the equilibrium has been reached after that.

A state that is updated only once from a state that seems to have reached equilibrium is almost the same as the original state, so cannot be considered as an independent sample. Therefore, we keep updating and suppose we reach $k' \gg k$.[21] The state at that time can be regarded as an independent sample from the previous sample. We repeat this procedure.

Taking an average over the resulting set of the states means that, due to the law of large numbers, we can calculate the average for states of high probabilities in the equilibrium distribution. In other words, we can do the following:

$$\sum_i P_{\text{eq}}(s_i)O(s_i) \approx \frac{1}{N_{\text{smp}}} \sum_k^{\text{MC}} O(s_k) \,. \tag{5.67}$$

Here, k represents the steps of a long separated Markov chain (MC) as described above. And N_{smp} indicates how many states were used. Including the error of the deviation from the equilibrium distribution, we have

$$\sum_i P_{\text{eq}}(s_i)O(s_i) = \frac{1}{N_{\text{MC}}} \sum_k^{\text{MC}} O(s_k) + O\left(\frac{1}{\sqrt{N_{\text{MC}}}}\right) \tag{5.68}$$

It is essentially important that this error does not depend on the range of i on the left-hand side. Namely, no matter how large the dimensionality of state space is, if the states can be updated using Markov chains and many states can be obtained, the expectation value can be precisely evaluated. This method, used to randomly assemble the states that are likely to be realized with high probability using Markov chains, is called the **Markov chain Monte Carlo (MCMC) method**.

Finally, we make a comment on the origin of the condition $k \gg 1$. As we have seen, the convergence destination is unique due to the Perron-Frobenius theorem for the transition matrix $\mathbb{T}$. And the convergence destination is the eigenvector itself for the maximum eigenvalue 1. The other way of saying it is that the eigenvector can be obtained from an appropriate initial vector by multiple matrix products,[22] and its convergence is determined by the second largest eigenvalue of $\mathbb{T}$.[23]

[20] The first k which satisfies $\mathbf{P}_{\text{eq}} \approx \mathbb{T}^k \mathbf{P}_0$ is called burn-in time or thermalization time.

[21] The autocorrelation time is a measure to give a suitable $|k' - k|$. See [67, 68] for details.

[22] This is so-called power method.

[23] Detailed discussions are found in [69].

In the next section, we introduce concrete Markov chain Monte Carlo algorithms.

5.3 Sampling Method with the Detailed Balance

In this section, as sampling methods that satisfy the detailed balance, we introduce the (narrowly defined) Metropolis method (algorithm) [70] and the heatbath method (algorithm) [71] (which is the Gibbs sampler in the machine learning context [72]). The method introduced here is generally called the Metropolis-Hastings algorithm [73]. Since these are derived from the principle of detailed balance described in the previous section, they converge to the target probability distribution.[24]

5.3.1 Metropolis Method

Here, we introduce the **Metropolis method** in a narrow sense, as a concrete method for how to take the transition probability $P(s_i|s_j)$. As a feature of the Metropolis method, there is no need to explicitly calculate the probability distribution, and it is sufficient if there is only a **Hamiltonian** (energy function) in physics.

To make the discussion concrete, consider the equilibrium distribution $P_{\text{eq}}(s_j)$ as follows:

$$P_{\text{eq}}(s_i) = \frac{1}{Z_\beta} e^{-\beta H[s_i]} . \tag{5.69}$$

Here, s_i is a state, and $H[s_i]$ is the Hamiltonian of the system.

Now, prepare two states s_i and s_j such that $H[s_i] < H[s_j]$. The state s_i has lower energy than the state s_j, so it is plausible that s_i is realized with high probability. From this consideration, let us take the transition probabilities as follows:

$$P(s_i|s_j) = 1 . \tag{5.70}$$

Then, the transition probability of the opposite direction $P(s_j|s_i)$ is determined from the principle of detailed balance,

$$e^{-\beta H[s_j]} = P(s_j|s_i) e^{-\beta H[s_i]} . \tag{5.71}$$

Namely, we get

$$P(s_j|s_i) = e^{-\beta(H[s_j] - H[s_i])} . \tag{5.72}$$

---

[24] We will not introduce global updates such as cluster algorithms, exchange Monte Carlo, or Hamiltonian (or hybrid) Monte Carlo (HMC) used in the context of Bayesian statistics. Interested readers should also learn about these.

Note that the transition probability depends only on the difference of the energy. In other words, if we change the state by some appropriate method and look at the change in energy, we will know the transition probability. Since it is not necessary to find the entire transition matrix (or the partition function), it is a versatile algorithm that can be applied regardless of the number of states, if we do not question if it is an optimal one.

The steps to execute the Metropolis method are summarized as follows:

1. First, prepare an appropriate state s_0. Hereinafter, repeat $i = 0, 1, 2, 3 \cdots$.
2. Calculate Hamiltonian ($= H[s_i]$) at s_i.
3. Change the state s_i to $s_{\text{candidate}}$ by some means.
4. Calculate Hamiltonian $= H[s_{\text{candidate}}]$ of $s_{\text{candidate}}$ after the change.
5. If $H[s_{\text{candidate}}]$ is smaller than $H[s_i]$, proceed as $s_{i+1} = s_{\text{candidate}}$. This is called *accepted*.
   Otherwise, if $e^{-\beta(H[s_{i+1}]-H[s_i])}$ is smaller than the random number $0 < r < 1$, accept $s_{\text{candidate}}$. If all the above conditions are not met, $s_{i+1} = s_i$. (This is called *rejected*.)

The last part (step 5) is also called the **Metropolis test**. β is the inverse temperature in statistical mechanics, but in general calculations, βH may be redefined as H and renormalized.

At this stage, how to update the state has not yet been specified, and it depends on the system. As an example of the update, in the case of the Ising model, one can try reversing the spin of one site.[25]

5.3.2 Heatbath Method

In physics, what is called the **heatbath method** can also be derived from the principle of fine balance. In the field of machine learning, it is also called the Gibbs sampler. The heatbath method focuses on a single degree of freedom in the system, divides the whole system into the single degree of freedom and the other degrees (heatbath), and uses only the probability of existence of the single degree of freedom in contact with the heatbath (irrespective of that extracted single degree of freedom), and determines the next state of that degree of freedom.

Let us assume that the transition probability from state s_i to state s_j is proportional to the existence probability of the state s_j. That is, we take it as $P(s_j|s_i) \propto P_{\text{eq}}(s_j)$. Including the normalization factor, we have

$$P(s_j|s_i) = \frac{P_{\text{eq}}(s_j)}{\sum_k P_{\text{eq}}(s_k)}. \tag{5.73}$$

[25]The HMC (Hamiltonian/Hybrid Monte Carlo) method can be regarded as one example of the Metropolis method, although we will not introduce it in this book.

The denominator would be an integral if the state were continuous. Since this form is the principle of detailed balance itself, it is always possible to update the states while satisfying the principle of detailed balance. The heatbath method also has the feature that, by definition, updating is not dependent on the current state. On the other hand, it is necessary to calculate the transition probabilities to all states, or in a physics terminology the local partition function (local free energy). In the case of a physical system, this is feasible if the Hamiltonian is written as a local sum as will be described below. In this case, the nearest neighbors connected to the degree of freedom (a single spin in the Ising model) of interest are regarded as a heat bath, which is the origin of the term.

As a specific example, we explain the heatbath method for the **Ising model**. The Hamiltonian of the Ising model is, on site i,

$$H_i = -S_i \sum_{j \in \langle i,j \rangle} S_j + (\text{terms not related to } i). \tag{5.74}$$

If we take the sum over all i, it gives the Hamiltonian of the whole system. Here, S_i is the Ising spin, and $\langle i, j \rangle$ means the set of the points of the nearest neighbor of i. In the case of the Ising model, we can write the Hamiltonian for each site in this way, and also write down the local Boltzmann weight $\exp(-\beta H_i)$ accompanying it. The value that the spin S_i at the site i can take in the next step can be determined as follows. From the definition of the transition probability for one spin at site i, we find, for any state $*$,

$$P_i(+|*) = \frac{\exp[-\beta R_i]}{\exp[-\beta R_i] + \exp[\beta R_i]} \quad \text{(the spin becomes } +1 \text{ at the next step)}, \tag{5.75}$$

$$P_i(-|*) = \frac{\exp[\beta R_i]}{\exp[-\beta R_i] + \exp[\beta R_i]} \quad \text{(the spin becomes } -1 \text{ at the next step)}. \tag{5.76}$$

Here, the energy contribution from the nearest neighbor is $R_i = -\sum_{j \in <i,j>} S_j$. A closer look, $P_i(+|*) = 1 - P_i(-|*)$, can further simplify the expression.

To summarize the above steps, we define

$$\omega^{(i)} = \frac{\exp[-\beta R_i]}{\exp[-\beta R_i] + \exp[\beta R_i]}, \tag{5.77}$$

then we use the uniform random number $\xi \in [0, 1)$ to perform the update by the rule

$$S_i^{\text{next}} = \begin{cases} 1 & (\xi < \omega^{(i)}), \\ -1 & \text{other than that.} \end{cases} \tag{5.78}$$

This is the heatbath method for the Ising model. We see that the spins other than the spin of our focus are considered as a heatbath and fixed. Also, thanks to focusing on one site, we have only two states, and the denominator of the transition probability can be calculated. We perform this for each point and update the whole.[26] However, because of this framework, it is not always possible to use the heatbath method for general models.

Column: From Ising Model to Hopfield Model

Let us take a look at the **Ising model**, an important model that connects machine learning and physics, and related models.[27] The Ising model is a model that explains a magnet from a microscopic point of view, with a variable (spin) that can take ± 1 on each lattice site. We call this a configuration and write $s = \{s_n\}_n$, where n is the coordinate specifying the lattice site. The **Hamiltonian** $H[s]$ of the Ising model is a function of the configuration of spins. Using Hamiltonian of this Ising model, the **partition function** Z_β is written as follows:

$$Z_\beta = \sum_{\{s\}} e^{-\beta H[s]}, \tag{5.79}$$

where $\beta = 1/(k_B T)$ is the inverse temperature. The distribution function is a function with respect to temperature. The sum for this $\{s\}$ is the sum for all possible spin configurations. Distribution functions are formally written in the form of sums, but the sums are generally not calculable. Currently, the summation can be done only in 1D and 2D Ising models [74, 75]. This is a common problem not only in the Ising model, but also in many systems dealt with by statistical mechanics.

Let us consider a 1D Ising model to get an insight here. The one-dimensional Ising model is a model of a magnet developed by W. Lenz and solved by his student, E. Ising, and is given by the following Hamiltonian:

$$H[s] = -J \sum_{\langle i,j \rangle} s_i s_j \,. \tag{5.80}$$

Here i and j are integers which can take values $1 \cdots, L$, where the size of the system is L. The value of the spin is $s_i = \pm 1$. $J(>0)$ is a coupling constant, which we take to be $J = 1$ in the following. In this case, it is called a ferromagnetic Ising model because the energy decreases when the signs of the spins are aligned. Also,

[26] Since the even-numbered points are not closest to each other, they can be updated at the same time, so that part can be parallelized.

[27] Here, we use the formulation of statistical mechanics without explaining the statistical mechanics itself.

$\langle i, j \rangle$ is a symbol that indicates that the sum is taken for the **nearest neighbor** pairs of the lattice points.

If we consider all possible configurations for the case where the length of the system is $L = 3$, there are eight possible configurations:

$$\{s\} = \uparrow\uparrow\uparrow \quad H[s] = -3 \quad \sum_j s_j = 3,$$

$$\{s\} = \uparrow\uparrow\downarrow \quad H[s] = 1 \quad \sum_j s_j = 1,$$

$$\{s\} = \uparrow\downarrow\uparrow \quad H[s] = 1 \quad \sum_j s_j = 1,$$

$$\{s\} = \uparrow\downarrow\downarrow \quad H[s] = 1 \quad \sum_j s_j = -1,$$

$$\{s\} = \downarrow\uparrow\uparrow \quad H[s] = 1 \quad \sum_j s_j = 1,$$

$$\{s\} = \downarrow\uparrow\downarrow \quad H[s] = 1 \quad \sum_j s_j = -1,$$

$$\{s\} = \downarrow\downarrow\uparrow \quad H[s] = 1 \quad \sum_j s_j = -1,$$

$$\{s\} = \downarrow\downarrow\downarrow \quad H[s] = -3 \quad \sum_j s_j = -3.$$

We also show the sum of the values of the Hamiltonian $H[s]$ and the spin ($\propto$ magnetization) for the periodic boundary condition. For a generic L, there are 2^L spin configurations, and we find that it is difficult to add up all of them. We will not go into details here, but it can be solved exactly, using a method called the transfer matrix method. Namely, the closed function form of Z_β can be obtained. From the exact solution, we know that the 1D Ising model cannot be ferromagnetic at any low temperature, but is paramagnetic.

In classical statistical mechanics, realization of any configuration is subject to probabilities.[28] The probability for realizing a spin configuration s at the inverse temperature β is given by

$$P(s) = \frac{1}{Z_\beta} e^{-\beta H[s]}. \tag{5.81}$$

---

[28] In quantum statistical mechanics for fermions, there are examples where the probability interpretation does not hold, such as a negative sign problem.

Using this realization probability, the expectation values are defined. For example, the **expectation value** of the spatial average of the spins is

$$\langle M \rangle = \lim_{V \to \infty} \frac{1}{V} \sum_{\{s\}} \sum_{n} P(s)\sigma_n \,. \tag{5.82}$$

Here V is the spatial volume.[29] The Helmholtz **free energy** is

$$F(\beta) = -\frac{1}{\beta} \log Z_\beta \,, \tag{5.83}$$

and the other physical quantities are obtained from the derivative of the Helmholtz free energy. Once a closed form of the Helmholtz free energy is obtained, then we say that the model is solved as a statistical mechanics.

Major ways to solve the Ising model is as follows:

1. Mean field approximation
2. Transfer matrix method

The mean field approximation is a self-consistent approximation method in which one equates the magnetic field generated by magnetization with the external magnetic field. In the case of one dimension, it gives a wrong answer, but in two dimensions or more, it gives a qualitatively correct result. The transfer matrix method is an effective method for one and two dimensions and gives exact solutions. However, as of January 2019, no solution to the Ising model in more than two dimensions is known.

As a model with a non-trivial phase structure, we consider a two-dimensional Ising model. The two-dimensional Ising model is given by the following Hamiltonian:

$$H[s] = -J \sum_{\langle \mathbf{i}, \mathbf{j} \rangle} s_{\mathbf{i}} s_{\mathbf{j}} \,. \tag{5.84}$$

Here $\mathbf{i} = (i_x, i_y)$ is a point on the lattice, and $\langle \mathbf{i}, \mathbf{j} \rangle$ represents the nearest neighbor pair. If the size of the system is $L_x \times L_y$, the number of states is $2^{L_x \times L_y}$. It is known that this model can also be solved by the transfer matrix method. From the exact solution, we know that there exists a phase transition at a finite temperature $T_c = \frac{2}{\log(\sqrt{2}+1)}$ [74, 75], and that is a transition between paramagnetic and ferromagnetic phases. In Chap. 8, we will explain to use a neural network to detect this phase transition.

[29] Strictly speaking, in order to determine the ground state to be unique, we need to break the symmetry by turning on the external field once, and after taking the limit of infinite volume, we then take the limit of the vanishing external field, and calculate $\langle M \rangle$.

By the way, by extending the Ising model, we will find the Hamiltonian of the **Hopfield model** which appears in the context of machine learning (see Chap. 10). First, consider the Edwards–Anderson model, which is a slightly extended version of the Ising model. The coupling constant of the nearest neighbors depends on the location of the spins,

$$H[s] = -\sum_{\langle i,j \rangle} k_{i,j} s_i s_j - B \sum_i s_i \,. \tag{5.85}$$

When viewed as a spinglass model, the Gaussian mean is typically taken for the coupling constant after taking the state sum for s. Not only can we consider the nearest neighbor dependence of the coupling constant, we can also consider a model with a coordinate-dependent magnetic field,

$$H[s] = -\sum_{i,j} k_{i,j} s_i s_j - \sum_i B_i s_i \,. \tag{5.86}$$

This is called the energy function of the Hopfield model.[30] This is the Hamiltonian used for Boltzmann machines. This form of Hamiltonian also appears in the description of spinglass models.

Here we describe the difference between the viewpoints of statistical mechanics and machine learning. In statistical mechanics, the magnetization $\langle M \rangle$, which is the spatial average of spins, for a given inverse temperature β and a given external magnetic field B, is obtained. On the other hand, in machine learning, we have to find $k_{i,j}$ that reproduces a given s, which is exactly an inverse problem. These inverse problems are generally described in Chap. 7.

Now, let us double the number of species of spins and call them v and h,

$$H[v,h] = -\sum_{i,j} k_{i,j}^{vv} v_i v_j - \sum_i B_i^v v_i - \sum_{i,j} k_{i,j}^{hh} h_i h_j - \sum_i B_i^h h_i - \sum_{i,j} k_{i,j}^{vh} v_i h_j \,. \tag{5.87}$$

k^{vv} is the coupling between v spins, k^{hh} is the coupling between h spins, k^{vh} is the coupling between the spin v and the spin h. B is the coupling of the external field to each spin. This model also has a name, the **Boltzmann machine**, and will appear in the next chapter. To avoid some difficulties, the restricted Boltzmann machine with $k_{i,j}^{vv} = k_{i,j}^{hh} = 0$ is actually used. The details will be explained in the next chapter. In this way, although the Ising model is a simple model, it has become a source of ideas and applications for machine learning.

---

[30]Strictly speaking, it is necessary to impose Hebb's rule and a symmetry on the coupling constant $k_{i,j}$, but we do not get into the details here.

Chapter 6
Unsupervised Deep Learning

Abstract In this chapter, at the end of Part I, we will explain Boltzmann machines and Generative adversarial networks (GANs). Both models are not the "find an answer" network given in Chap. 3, but rather the network itself giving the probability distribution of the input. Boltzmann machines have historically been the cornerstone of neural networks and are given by the Hamiltonian statistical mechanics of multi-particle spin systems. It is an important bridge between machine learning and physics. Generative adversarial networks are also one of the important topics in deep learning in recent years, and we try to provide an explanation of it from a physical point of view.

Chapters 2 and 3 described machine learning when a data set consisting of an input and a training label is given. In this chapter, we change the subject and explain the method of machine learning when no training label is given: the case of a generative model.[1]

6.1 Unsupervised Learning

What is **unsupervised learning**? It is a machine learning scheme that uses data that has only "input data" and no "answer signal" for it:

$$\{\mathbf{x}[i]\}_{i=1,2,\ldots,\#}\,. \tag{6.1}$$

[1]In addition to generative models, tasks such as clustering and principal component analysis are included in unsupervised learning. It should be noted here that in the sense that data is provided in advance, it is different from schemes such as **reinforcement learning** where data is not provided. Reinforcement learning is a very important machine learning technique that is not discussed in this book. Instead, here we provide some references. First, the standard textbook is Ref. [76] written by Sutton et al. This also describes the historical background.

A. Tanaka et al., *Deep Learning and Physics*, Mathematical Physics Studies,
https://doi.org/10.1007/978-981-33-6108-9_6

Even in this case, we assume that there exists some probability distribution $P(\mathbf{x})$,

$$\mathbf{x}[i] \sim P(\mathbf{x}) . \tag{6.2}$$

So now, let us consider that there is a model $Q_J(\mathbf{x})$, that is, a probability distribution with a set of parameters J, and we want to make it closer to $P(\mathbf{x})$. In other words, we want to reduce

$$K(J) = \int d\mathbf{x}\; P(\mathbf{x}) \log \frac{P(\mathbf{x})}{Q_J(\mathbf{x})} . \tag{6.3}$$

Once this minimization is achieved, $Q_J(\mathbf{x})$ can be used to create "fake" data.

6.2 Boltzmann Machine

First, as usual, we consider a "statistical mechanics" model as $Q_J(\mathbf{x})$. For example, consider a Hamiltonian with an interaction for each component of $\mathbf{x}$ as

$$H_J(\mathbf{x}) = \sum_i x_i J_i + \sum_{ij} x_i J_{ij} x_j + \dots \tag{6.4}$$

and the model is given as

$$Q_J(\mathbf{x}) = \frac{e^{-H_J(\mathbf{x})}}{Z_J} . \tag{6.5}$$

It is only necessary to adjust the "coupling constant" J to reduce (6.3). This is called a Boltzmann machine. The simplest algorithm would be to differentiate (6.3) and use the derivative to change the values of J:

$$J \leftarrow J - \epsilon \partial_J K(J). \tag{6.6}$$

The derivative is given by

$$\begin{aligned} \partial_J K(J) &= \partial_J \int d\mathbf{x}\; P(\mathbf{x}) \log \frac{P(\mathbf{x})}{Q_J(\mathbf{x})} \\ &= - \int d\mathbf{x}\; P(\mathbf{x}) \partial_J \log Q_J(\mathbf{x}) \\ &= - \int d\mathbf{x}\; P(\mathbf{x}) \partial_J \Big[- H_J(\mathbf{x}) - \log Z_J \Big] \\ &= \langle \partial_J H_J(\mathbf{x}) \rangle_P - \langle \partial_J H_J(\mathbf{x}) \rangle_{Q_J} . \end{aligned} \tag{6.7}$$

Here

$$\langle \bullet(\mathbf{x}) \rangle_P = \int d\mathbf{x}\, P(\mathbf{x}) \bullet (\mathbf{x}) \tag{6.8}$$

represents the expectation value by the probability distribution P.

Training in actual cases
In reality, $P(\mathbf{x})$ is not directly known, but what is known is the data (6.1) which is regarded as a **sampling** from that. In this case, the first term in (6.7) can be approximated by replacing the integral with the sampling average. On the other hand, it is difficult to calculate the second term even if the form of $Q_J(\mathbf{x})$ is known, because it is necessary to calculate the partition function.[2] So usually, we approximate also the second term by some kind of sampling using a model,

$$\mathbf{y}[i] \sim Q_J(\mathbf{x}) \tag{6.9}$$

to obtain

$$\begin{aligned} \partial_J K(J) &= \langle \partial_J H_J(\mathbf{x}) \rangle_P - \langle \partial_J H_J(\mathbf{x}) \rangle_{Q_J} \\ &\approx \sum_{i=1}^{N_{\text{positive}}} \frac{1}{N_{\text{positive}}} \partial_J H_J(\mathbf{x}[i]) - \sum_{i=1}^{N_{\text{negative}}} \frac{1}{N_{\text{negative}}} \partial_J H_J(\mathbf{y}[i]) \,. \end{aligned} \tag{6.10}$$

Although the approximation of the first term in (6.7) cannot be improved, the quality of the algorithm will depend on how well the sampling in the second term is performed.

6.2.1 *Restricted Boltzmann Machine*

In fact, with a little ingenuity, a Boltzmann machine with a slightly more efficient learning algorithm can be constructed. To do this, we introduce a "hidden degree of freedom" $\mathbf{h}$ and consider the following Hamiltonian:

$$H_J(\mathbf{x}, \mathbf{h}) = \sum_i x_i J_i + \sum_\alpha h_\alpha J_\alpha + \sum_{i\alpha} x_i J_{i\alpha} h_\alpha \tag{6.11}$$

[2] It is very difficult to calculate the partition function using the Monte Carlo method. Roughly speaking, the partition function needs information on all states, but the Monte Carlo method focuses on information on parts that contribute to expectation values.

and the resultant **effective Hamiltonian**,[3]

$$H_J^{\text{eff}}(\mathbf{x}) = \log Z_J - \log \sum_{\mathbf{h}} e^{-H_J(\mathbf{x},\mathbf{h})} . \tag{6.12}$$

And we define the model as

$$Q_J(\mathbf{x}) = e^{-H_J^{\text{eff}}(\mathbf{x})} . \tag{6.13}$$

The model whose interaction is restricted in this way is called **restricted Boltzmann machine**.

Learning in a restricted Boltzmann machine
Substituting (6.12) for $H_J(\mathbf{x})$ in the expression (6.10), we find

$$\begin{aligned}
&\partial_J K(J) \\
&= \langle \partial_J H_J^{\text{eff}}(\mathbf{x}) \rangle_{\mathbf{x}\sim P} - \langle \partial_J H_J^{\text{eff}}(\mathbf{y}) \rangle_{\mathbf{y}\sim Q_J} \\
&= \langle \partial_J \Big[\log Z_J - \log \sum_{\mathbf{h}} e^{-H_J(\mathbf{x},\mathbf{h})} \Big] \rangle_{\mathbf{x}\sim P} - \langle \partial_J \Big[\log Z_J - \log \sum_{\mathbf{h}} e^{-H_J(\mathbf{y},\mathbf{h})} \Big] \rangle_{\mathbf{y}\sim Q_J} \\
&= \left\langle \frac{\sum_{\mathbf{h}_n} e^{-H_J(\mathbf{x},\mathbf{h}_n)} \partial_J H_J(\mathbf{x},\mathbf{h}_n)}{\sum_{\mathbf{h}_d} e^{-H_J(\mathbf{x},\mathbf{h}_d)}} \right\rangle_{\mathbf{x}\sim P} - \left\langle \frac{\sum_{\mathbf{h}_n} e^{-H_J(\mathbf{y},\mathbf{h}_n)} \partial_J H_J(\mathbf{y},\mathbf{h}_n)}{\sum_{\mathbf{h}_d} e^{-H_J(\mathbf{y},\mathbf{h}_d)}} \right\rangle_{\mathbf{y}\sim Q_J} \\
&= \left\langle \sum_{\mathbf{h}_n} P_J(\mathbf{h}_n|\mathbf{x}) \partial_J H_J(\mathbf{x},\mathbf{h}_n) \right\rangle_{\mathbf{x}\sim P} - \left\langle \sum_{\mathbf{h}_n} P_J(\mathbf{h}_n|\mathbf{y}) \partial_J H_J(\mathbf{y},\mathbf{h}_n) \right\rangle_{\mathbf{y}\sim Q_J} .
\end{aligned} \tag{6.14}$$

Here

$$P_J(\mathbf{h}_n|\mathbf{x}) = \frac{e^{-H_J(\mathbf{x},\mathbf{h}_n)}}{\sum_{\mathbf{h}_d} e^{-H_J(\mathbf{x},\mathbf{h}_d)}} \tag{6.15}$$

is the conditional probability described in the column in Chap. 2. The inverse conditional probability

$$P_J(\mathbf{x}_n|\mathbf{h}) = \frac{e^{-H_J(\mathbf{x}_n,\mathbf{h})}}{\sum_{\mathbf{x}_d} e^{-H_J(\mathbf{x}_d,\mathbf{h})}} \tag{6.16}$$

will also be needed below.

[3] The effective Hamiltonian is defined to satisfy the following equation: $\exp[-H_J^{\text{eff}}(\mathbf{x})] = \sum_{\mathbf{h}} \exp[-H_J(\mathbf{x},\mathbf{h})]/Z_J$.

Training of a restricted Boltzmann machine

At this rate, $\mathbf{y} \sim Q_J(\mathbf{y})$ is needed to approximate the second term of (6.14), and so the partition function needs to be calculated, just as in the case of the normal Boltzmann machine. To improve the situation, in the training of the restricted Boltzmann machine, the sampling $\mathbf{y} \sim Q_J(\mathbf{y})$ of our concern is replaced by

$$\mathbf{y} \sim P_J(\mathbf{x}_n|\mathbf{h}), \quad \mathbf{h} \sim P_J(\mathbf{h}|\mathbf{x}), \quad \mathbf{x} \sim P(\mathbf{x}). \tag{6.17}$$

This is called **contrastive divergence method**.[4] So the learning algorithm is

$$J \leftarrow J - \epsilon \delta J, \tag{6.18}$$

$$\delta J = \Big\langle \partial_J H_J(\mathbf{x}, \mathbf{h}_n) \Big\rangle_{\mathbf{h}_n \sim P_J(\mathbf{h}_n|\mathbf{x}),\ \mathbf{x} \sim P(\mathbf{x})} - \Big\langle \partial_J H_J(\mathbf{y}, \mathbf{h}_2) \Big\rangle_{\mathbf{h}_2 \sim P_J(\mathbf{h}_2|\mathbf{y}),\ \mathbf{y} \sim P_J(\mathbf{x}_n|\mathbf{h}_1),\ \mathbf{h}_1 \sim P_J(\mathbf{h}_1|\mathbf{x}),\ \mathbf{x} \sim P(\mathbf{x})}. \tag{6.19}$$

Since this algorithm uses only the conditional probabilities (6.15) and (6.16) at the time of sampling, there is no need to calculate the first term of the partition function (6.12). This makes it a very fast learning algorithm. This method was proposed by Hinton in Ref. [77], and an explanation of why such a substitution could be made was given in Ref. [78] and other references. In the following, we provide a physical explanation of it.

Heatbath method and detailed balance

Once you have trained the restricted Boltzmann machine and obtained good parameters J^*, you can use it as a data sampling machine. The sampling is then done using the algorithm of the heatbath method as follows:

1. Initialize $\mathbf{x}$ appropriately and call it $\mathbf{x}_0$. (6.20)
2. Repeat the following for $t = 1, 2, \ldots$: (6.21)

$$\mathbf{h}_t \sim P_{J^*}(\mathbf{h}|\mathbf{x}_t), \tag{6.22}$$

$$\mathbf{x}_{t+1} \sim P_{J^*}(\mathbf{x}|\mathbf{h}_t). \tag{6.23}$$

In this sampling, the **transition probability** is written as

$$P_{J^*}(\mathbf{x}'|\mathbf{x}) = \sum_{\mathbf{h}} P_{J^*}(\mathbf{x}'|\mathbf{h}) P_{J^*}(\mathbf{h}|\mathbf{x}). \tag{6.24}$$

[4]The contrastive divergence method is abbreviated as "CD method." What is used here is also called the CD-1 method. In general, the contrastive divergence method is called the CD-k method, where k is the number of samples using the heatbath method for the process between $\mathbf{x}$ and $\mathbf{h}$.

Let us consider what is the **detailed balance** condition that this transition probability satisfies. To do that, we just need to know the difference between the original probability and the one where $\mathbf{x}$ and $\mathbf{x}'$ have been swapped in (6.24). As a trial, we write (6.24) with $\mathbf{x}$ and $\mathbf{x}'$ replaced, and try to get as close as possible to the original form:

$$P_{J^*}(\mathbf{x}|\mathbf{x}') = \sum_{\mathbf{h}} P_{J^*}(\mathbf{x}|\mathbf{h}) P_{J^*}(\mathbf{h}|\mathbf{x}') \\ = \sum_{\mathbf{h}} P_{J^*}(\mathbf{h}|\mathbf{x}') P_{J^*}(\mathbf{x}|\mathbf{h}) . \tag{6.25}$$

From the first line to the second line, the order of multiplication in the sum was changed. Of course, this is not equal to the original (6.24), but we notice that the conditional probability argument has been swapped. Using Bayes' theorem explained in the column of Chap. 2, the following identity holds:

$$P_J(\mathbf{h}|\mathbf{x}) e^{-H_J^{\text{eff}}(\mathbf{x})} = P_J(\mathbf{x}|\mathbf{h}) e^{-H_J^{\text{eff}}(\mathbf{h})} . \tag{6.26}$$

Here $H_J^{\text{eff}}(\mathbf{h})$ is the effective Hamiltonian for the hidden degrees of freedom,

$$H_J^{\text{eff}}(\mathbf{h}) = \log Z_J - \log \sum_{\mathbf{x}} e^{-H_J(\mathbf{x},\mathbf{h})} . \tag{6.27}$$

Using Bayes' theorem (6.26) to transform the two conditional probabilities of (6.25), we find that the terms $H_J^{\text{eff}}(\mathbf{h})$ just cancel each other and

$$(6.25) = \frac{e^{-H_{J*}^{\text{eff}}(\mathbf{x})}}{e^{-H_{J*}^{\text{eff}}(\mathbf{x}')}} P_{J^*}(\mathbf{x}'|\mathbf{x}) . \tag{6.28}$$

This is the equation of the detailed balance condition whose convergence limit is

$$Q_{J^*}(\mathbf{x}) = e^{-H_{J*}^{\text{eff}}(\mathbf{x})} . \tag{6.29}$$

So, if Q_{J^*} is sufficiently close to the target distribution P as in the scenario up to this point, it can produce a fake sample which is quite close to the real thing by sampling using the heatbath method. In addition, the contrastive divergence method is, in a sense, a "self-consistent"İ optimization. The reason is that the second term of the formula (6.19) was originally a sampling from Q_J, but if the training progresses and Q_J gets close enough to P, the second term is nothing but an iterative part of the heatbath algorithm described above. In fact, the derivation of the contrastive divergence method can be discussed from the point of view of such detailed balance.

6.3 Generative Adversarial Network

Unsupervised learning explained so far has basically been based on algorithms that reduce the relative entropy (6.3). However, this is not the only way to get the probability distribution Q_J closer to P. Here, we shall explain models called **generative adversarial networks, GANs** [79], which have been attracting attention in recent years.

Basic settings
In a GAN setting, we prepare two types of networks. In the following, the space where the values of each pixel of the image are arranged (the space where the data lives) is called X, and another space (called the latent space or the feature space) that the user of the GAN sets is called Z. Each network is a function

$$G : Z \to X, \tag{6.30}$$

$$D : X \to \mathbb{R}, \tag{6.31}$$

which is represented by a neural network. G is called a generator, and D is called a discriminator. As an analogy, G is counterfeiting, aiming to create as much elaborate and realistic data x_{fake} as possible, while D is a police officer and its learning objective is to be able to distinguish between real samples x_{real} and fake samples x_{fake}. The goal of GANs is to have two different (but adversarial) networks compete with each other to get G, which can produce fake data x_{fake} that can be mistaken for the real thing.

Probability distribution induced by G
GAN formulation is also based on probability theory. First, we set a probability distribution on Z by hand. This is a "seed" of the fake data, and we often take a distribution that is relatively easy to sample, such as a Gaussian distribution, a uniform distribution on a spherical surface,[5] or a uniform distribution in a box $[-1, 1]^{\dim Z}$. We name a chosen distribution $p_z(z)$. G takes the "seed" z as an argument and transfers it to a point in the data space, and the probability distribution on X is induced from p_z,

$$Q_G(x) = \int dz \; p_z(z)\delta\Big(x - G(z)\Big) . \tag{6.32}$$

Here δ is the Dirac delta function often used in quantum mechanics and electromagnetism. In other words, this is the same as

$$x \sim Q_G(x) \Leftrightarrow z \sim p_z(z), x = G(z) . \tag{6.33}$$

[5]By the way, since Z usually brings a space of several hundred dimensions, there is not much difference between the Gaussian distribution and the uniform distribution on the sphere due to the effect of the curse of dimensionality. This is because the higher the dimension, the larger the ratio of the spherical shell to the inside of the spherical surface.

Objective function and learning process

The interesting point about GANs is that it does not use the relative entropy (6.3) directly to optimize (6.32). Instead, one considers the following function:[6]

$$V_D(G, D) = \langle \log(1 + e^{-D(x)}) \rangle_{x \sim P(x)} + \langle \log(1 + e^{+D(x)}) \rangle_{x \sim Q_G(x)}, \tag{6.35}$$

$$V_G(G, D) = -V_D(G, D). \tag{6.36}$$

This just means: (the value of V_D is small) $\Leftrightarrow$ ($D(x_{\text{real}})$ =large, $D(x_{\text{fake}})$ = small). If we train D to reduce V_D, then we can put a role as a "police officer" to D. On the other hand, according to (6.36), decreasing V_G = increasing V_D, so if this value of V_G is reduced, G can trick D. Namely, the learning process of GANs is to repeat the following updates:

$$G \leftarrow G - \epsilon \nabla_G V_G(G, D), \tag{6.37}$$

$$D \leftarrow D - \epsilon \nabla_G V_D(G, D), \tag{6.38}$$

This optimization is a search for a Nash equilibrium,

$$G^* \ s.t. \forall G, \ V_G(G^*, D^*) \leq V_G(G, D^*), \tag{6.39}$$

$$D^* \ s.t. \forall D, \ V_D(G^*, D^*) \leq V_D(G^*, D). \tag{6.40}$$

In fact, with the condition of (6.36), that is, the sum of the objective function of G and the objective function of D becomes zero, the equilibrium point is shown to satisfy

$$V_D(G^*, D^*) = \min_G \max_D V_D(G, D). \tag{6.41}$$

This is a consequence of the minimax theorem [80] by von Neumann.[7] From this condition, it is relatively easy to prove why Q_G reproduces P. First, we look at max_D of (6.41). D is a function on X and the functional V_D is differentiable; this

[6]This is written with the sigmoid function $\sigma(u) = (1 + e^{-u})^{-1}$ as

$$-\langle \log \sigma(D(x)) \rangle_{x \sim P(x)} - \langle \log(1 - \sigma(D(x))) \rangle_{x \sim Q_G(x)} \tag{6.34}$$

and it is similar to the cross-entropy error (3.16), which is the error function derived for binary classification in Chap. 3. In fact, this is identical to a binary classification problem of discriminating whether an item of data is real ($x \sim P(x)$) or not ($x \sim Q_G(x)$).

[7]The minimax theorem is that if $f(x, y)$ is a concave (convex) function with respect to the 1st (2nd) variable x (y),

$$\min_x \max_y f(x, y) = \max_y \min_x f(x, y) \,. \tag{6.42}$$

Taking $f = V_D$ and setting the solution of this minimax problem to $\overline{G}, \overline{D}$, one can prove that these satisfy the Nash equilibrium condition (6.39) and (6.40). In the proof, one uses the condition $V_G + V_D = 0$ (the **zero-sum condition**).

can be regarded as a variational problem in physics, to seek for a solution of the equation

$$0 = \frac{\delta V_D(G, D)}{\delta D(x)} . \tag{6.43}$$

The variation of V_D is

$$\begin{aligned} \delta V_D(G, D) &= \delta \int dx \Big(P(x) \log(1 + e^{-D(x)}) + Q_G(x) \log(1 + e^{+D(x)}) \Big) \\ &= \int dx\, \delta D(x) \Big(- P(x) \frac{e^{-D(x)}}{1 + e^{-D(x)}} + Q_G(x) \frac{e^{+D(x)}}{1 + e^{+D(x)}} \Big) . \end{aligned} \tag{6.44}$$

So, the solution $D = D^*$ makes the quantity in the parentheses vanish, which leads to

$$e^{-D^*(x)} = \frac{Q_G(x)}{P(x)} . \tag{6.45}$$

Substituting this into $V_G = -V_D$ gives the objective function of $\min_G$,

$$\begin{aligned} V_G(G, D^*) &= \int dx \Big(P(x) \log \frac{P(x)}{P(x) + Q_G(x)} + Q_G(x) \log \frac{Q_G(x)}{P(x) + Q_G(x)} \Big) \\ &= D_{KL}\Big(P \Big|\Big| \frac{P + Q_G}{2} \Big) + D_{KL}\Big(Q_G \Big|\Big| \frac{P + Q_G}{2} \Big) - 2 \log 2 . \end{aligned} \tag{6.46}$$

Due to the property of relative entropy, minimizing this functional with respect to G has to be attained by

$$P(x) = \frac{P(x) + Q_G(x)}{2} = Q_G(x) . \tag{6.47}$$

Training in actual cases

However, in actual GAN training, things do not go as in the theory, causing various learning instabilities. An early perceived problem was that D became too strong first, and the gradient for G disappeared. To avoid this, in most of the cases, instead of (6.36),

$$V_G(G, D) = \langle \log(1 + e^{-D(x)}) \rangle_{x \sim Q_G(x)} \tag{6.48}$$

is used. The above proof heavily relied on (6.41), so there is no theoretical guarantee that replacing V_G will work. The GAN has been recognized explosively since the announcement of implementing it to a convolutional neural network (deep convolutional GAN, DCGAN) [42] for learning more stably. From this, it is expected that creating a theory of why it works will require incorporating conditions not only about the algorithm but also about the structure of the actual network used. As of April 2019, to the best of the authors' knowledge, the theoretical guarantees of the (6.48) version of GANs are an open question.

6.3.1 Energy-Based GAN

Another example of $V_{G,D}$ is called an energy-based GAN [81], which is defined as

$$V_D(G, D) = \langle D(x)\rangle_{x\sim P(x)} + \langle \max(0, m - D(x))\rangle_{x\sim Q_G(x)}, \tag{6.49}$$

$$V_G(G, D) = \langle D(x)\rangle_{x\sim Q_G(x)}. \tag{6.50}$$

Here $m(> 0)$ is a manually set value that corresponds to the energy of the fake data. It is the energy to measure "how much data x looks real," which is modeled by the classifier D. As in an ordinary physical system, energy must be bounded from below:

$$D(x) \geq 0\,. \tag{6.51}$$

In the energy-based GAN, a classifier network is configured to satisfy this energy condition (6.51). Under these settings, The training of D is performed as follows:

$$\begin{array}{l}\text{If } x \text{ is real, make } D(x) \text{ close to 0.}\\ \text{If } x \text{ is fake, make } D(x) \text{ close to m.}\end{array} \tag{6.52}$$

This corresponds to making the value of (6.49) as small as possible.[8] The training of G is to make $x_{\text{fake}} = G(z)$ have the lowest possible energy (= realistic). The "equilibrium point" (6.39) and (6.40) based on (6.49) and (6.50) can be shown to give $Q_{G^*}(x) = P(x)$ without the help of the minimax theorem. The proof might be too mathematical, but here we shall show it.[9]

Property obtained from equilibrium point of V_D
The inequality (6.40) says that for $V_D(G^*, D)$ into which $G = G^*$ is substituted, D^* takes the minimum value. So we start by looking at what happens to $D(x)$ that minimizes the following:

$$V_D(G^*, D) = \int dx \Big[P(x)D(x) + Q_{G^*}(x)\max(0, m - D(x))\Big]. \tag{6.53}$$

[8]Just as V_D in the original GAN corresponds to the cross entropy, the objective function (6.49) called hinge loss corresponds to the error function of a support vector machine (which is not described in this book).

[9]This is a detailed version of the proof provided in the appendix of the original paper.

As before, one may want to take a variation with respect to $D(x)$, but unfortunately it does not work because of the condition (6.51) and the discontinuous function "max." So, we evaluate what the integrand looks like at each x. Then basically we just need to consider a one-variable function

$$f(D) = PD + Q\max(0, m - D), \quad 0 < P, Q < 1. \tag{6.54}$$

and we find that it is just a matter of division of cases:

$$\min f(D) = \begin{cases} f(D=0) = mQ \ (P < Q), \\ f(D=m) = mP \ (P \geq Q). \end{cases} \tag{6.55}$$

Then, there is this function for each point x, so we find

$$V_D(G^*, D^*) = m\Big[\int 1_{P(x)<Q_{G^*}(x)} dx\ P(x) + \int 1_{P(x)\geq Q_{G^*}(x)} dx\ Q_{G^*}(x)\Big]. \tag{6.56}$$

Here, $1_{\text{condition}}$ is a step function that takes 1 (0) when the condition is (not) satisfied. By definition

$$1_{P(x)\geq Q_{G^*}(x)} = 1 - 1_{P(x)<Q_{G^*}(x)}\,. \tag{6.57}$$

Substituting this into (6.56), using the fact that $Q_{G^*}(x)$ is a probability distribution and so its integration gives 1, we find

$$V_D(G^*, D^*) = m\Big[1 + \int 1_{P(x)<Q_{G^*}(x)} dx \Big(\underbrace{P(x) - Q_{G^*}(x)}_{<0}\Big)\Big] \leq m\,. \tag{6.58}$$

This inequality is the first important conclusion. The point here is that the relationship with the last m is $\leq$ instead of $<$. This is the heart of the proof, so let us elaborate on the explanation. The reason why we do not have $<$ is that there can be no[10] point x satisfying $P(x) - Q_{G^*}(x) < 0$. In this case, the contribution from the second term of (6.58) is zero due to the step function $1_{P(x)<Q_{G^*}(x)}$.

Property obtained from equilibrium point of V_G
Next, we write the inequality (6.39) by integration as

$$\forall G, \quad \int dx\ Q_{G^*}(x) D^*(x) \leq \int dx\ Q_G(x) D^*(x)\,. \tag{6.59}$$

[10] To be precise, it is better to say $P(x) - Q_{G^*}(x) \geq 0$ almost everywhere. The meaning of this expression is explained in the next footnote.

Can we extract any useful information from here? The point is to imagine a "perfect generator" G_{perfect},

$$Q_{G_{\text{perfect}}}(x) = P(x) . \tag{6.60}$$

Since in this proof G is assumed to have infinite expressive power (it can approximate all possible functions), we may substitute G_{perfect} into G. Since (6.59) holds for any G, the inequality holds also for the "perfect generator":

$$\int dx\ Q_{G^*}(x)D^*(x) \leq \int dx\ Q_{G_{\text{perfect}}}(x)D^*(x) = \int dx\ P(x)D^*(x) . \tag{6.61}$$

The integral in this last expression is the same as the first term of (6.53) into which $D = D^*$ is substituted, so using this inequality we newly obtain

$$V_D(G^*, D^*) \geq \int dx\ Q_{G^*}(x)\Big[D^*(x) + \max(0, m - D^*(x))\Big]. \tag{6.62}$$

Here, we can prove[11] that "almost everywhere"[12] the following holds:

$$D^*(x) \leq m . \tag{6.66}$$

[11]Let us prove it by reductio ad absurdum. The negation of the statement that almost everywhere we have (6.66) is

$$S = \{x|D^*(x) > m\}\text{contributes to the integral.} \tag{6.63}$$

We define a new $\tilde{D}(x) = \min(m, D^*(x))$ and substituting it into D of $V_D(G^*, D)$, we just put $D = \tilde{D}$ in (6.53),

$$\begin{aligned} V_D(G^*, \tilde{D}) &= \int_{S \cup S^c} dx\Big[P(x)\tilde{D}(x) + Q_{G^*}(x)\max(0, m - \tilde{D}(x))\Big] \\ &= \int_S dx\Big[P(x) \underbrace{\tilde{D}(x)}_{=m<D^*(x)} + Q_{G^*}(x)\underbrace{\max(0, m - \tilde{D}(x))}_{=0<\max(0,m-D^*(x))}\Big] \\ &\quad + \int_{S^c} dx\Big[P(x)\underbrace{\tilde{D}(x)}_{D^*(x)} + Q_{G^*}(x)\max(0, m - \underbrace{\tilde{D}(x)}_{D^*(x)})\Big] \\ &< \int_{S \cup S^c} dx\Big[P(x)D^*(x) + Q_{G^*}(x)\max(0, m - D^*(x))\Big] = V_D(G^*, D^*) . \end{aligned} \tag{6.64}$$

This inequality contradicts the Nash equilibrium definition (6.40), where D^* gives the minimum value of $V_D(G^*, D)$ for D.

[12]This is equivalent to

$$\int 1_{D^*(x)>m} dx = 0 . \tag{6.65}$$

In other words, the support set $S = \{x|D^*(x) > m\}$ does not contribute to the integral.

So we find

$$V_D(G^*, D^*) \geq \int dx\ Q_{G^*}(x)\Big[D^*(x) + \underbrace{\max(0, m - D^*(x))}_{=m-D^*(x)}\Big] = m. \tag{6.67}$$

This is the second important conclusion.

Proof completion
Finally, two inequalities (6.58) and (6.67) including equalities, which differ only in the directions, hold, hence we find

$$V_D(G^*, D^*) = m\,. \tag{6.68}$$

Together with

$$(6.58){:}\, V_D(G^*, D^*) = m\Big[1 + \int 1_{P(x)<Q_{G^*}(x)} dx\ \big(\underbrace{P(x) - Q_{G^*}(x)}_{<0}\big)\Big] \leq m\,, \tag{6.69}$$

this leads to

$$0 = \int 1_{P(x)<Q_{G^*}(x)} dx\ \big(\underbrace{P(x) - Q_{G^*}(x)}_{<0}\big)\,. \tag{6.70}$$

Since the $\underbrace{\quad}$ part contributes only with negative values, this means that there is no integration range in the first place,

$$0 = \int 1_{P(x)<Q_{G^*}(x)} dx\,. \tag{6.71}$$

This means that we need $P(x) \geq Q_{G^*}(x)$ almost everywhere. Considering the case where the inequality sign is reversed,

$$\begin{aligned}
&\int 1_{P(x)>Q_{G^*}(x)} dx\Big(\underbrace{P(x) - Q_{G^*}(x)}_{>0}\Big) \\
&= \int (1 - 1_{P(x)\leq Q_{G^*}(x)}) dx\Big(P(x) - Q_{G^*}(x)\Big) \\
&= \underbrace{\int dx\Big(P(x) - Q_{G^*}(x)\Big)}_{1-1=0} - \int \underbrace{1_{P(x)\leq Q_{G^*}(x)}}_{\text{Integrand is zero for “=”}} dx\Big(P(x) - Q_{G^*}(x)\Big) \\
&= -\int 1_{P(x)<Q_{G^*}(x)} dx\Big(P(x) - Q_{G^*}(x)\Big) \overset{(6.70)}{=} 0\,.
\end{aligned} \tag{6.72}$$

For the same reason, we find

$$0 = \int 1_{P(x) > Q_{G^*}(x)} dx, \tag{6.73}$$

and now, almost everywhere, $P(x) \leq Q_{G^*}(x)$. So after all, almost everywhere

$$P(x) = Q_{G^*}(x). \tag{6.74}$$

This is what we wanted to show.

6.3.2 Wasserstein GAN

Another interesting extension [82] of the GAN is possible by considering unsupervised learning in the context of optimal transport [83, 84]. Optimal transport treats minimization of transport energy, so the model can be interpreted physically. Regarding the probability distribution Q as a pile of sand, and the probability distribution P as a hole of the same volume dug in the ground, then the optimal transport energy is defined as the minimum energy consumed for transporting sand from the pile to the hole to make a flat surface. One of the typical optimal transport energies is what is called the Wasserstein distance. The definition is

$$D_W(P, Q) = \min_{\pi \in (6.77)} U(\pi), \tag{6.75}$$

$$U(\pi) = \langle E(x, y) \rangle_{(x,y) \sim \pi(x,y)}. \tag{6.76}$$

Here $E(x, y)$ represents the transportation cost energy between data points x, y. Typically, we take the distance between x, y for it. $\pi(x, y)$ is a joint probability distribution of how much to transport $y \to x$, which is assumed to satisfy

$$\int dx\, \pi(x, y) = Q(y), \quad \int dy\, \pi(x, y) = P(x). \tag{6.77}$$

$U(\pi)$ is the expectation value of the transportation cost energy using this probability distribution π, and in the analogy to thermodynamics we can think of it as "internal energy." If the transportation cost $E(x, y)$ satisfies the property of distance:

$$\begin{cases} 0 \leq D_W(P, Q), \\ 0 = D_W(P, Q) \Leftrightarrow \forall x, P(x) = Q(x), \end{cases} \tag{6.78}$$

it has properties as a substitute for the relative entropy.

Generalization to Helmholtz free energy
The interesting thing about this distance is that there is an equivalent expression which looks completely different. In order to derive it, once we put this system in the nonzero temperature, and consider, not the minimum value of the internal energy, but the minimum value of Helmholtz free energy.[13]

$$D_W^T(P, Q) = \min_{\pi \in (6.77)} \Big(U(\pi) - TS(\pi) \Big) . \tag{6.79}$$

Here, we adopt the following entropy[14] of the transportation plan π,

$$S(\pi) = \langle -\log \pi(x, y) + 1 \rangle_{(x,y)\sim\pi(x,y)} . \tag{6.80}$$

Because of the temperature, this does not actually satisfy the property of distance (6.78). However, in thermodynamics, free energy often gives a better perspective than internal energy, and accordingly it acquires good properties.[15]

Now, we want to determine the transportation plan π that minimizes this free energy, and a difficulty is how to take into account the constraint (6.77). We put the constraint into the optimization by using the Lagrange multiplier method. The expectation value is written in the form of an integral as

$$\begin{aligned} D_W^T(P, Q) = \min_{\pi} \max_{f,g} \Bigg(&\int dxdy\, \pi(x, y) \Big[E(x, y) + T \log \pi(x, y) - T \Big] \\ &+ \int dx\, f(x) \Big[P(x) - \int dy\, \pi(x, y) \Big] \\ &+ \int dy\, g(y) \Big[Q(y) - \int dx\, \pi(x, y) \Big] \Bigg) . \end{aligned} \tag{6.81}$$

We consider changing the order, as $\min_\pi \max_{f,g} = \max_{f,g} \min_\pi$. To find the minimum value over π first, it is sufficient to take the variation of π and set it to zero,

$$0 = E(x, y) + T \log \pi^*(x, y) - f(x) - g(y) . \tag{6.82}$$

Then we can substitute this solution and consider $\max_{f,g}$, to find another expression

$$D_W^T(P, Q) = \max_{f,g} \Big(\langle f(x) \rangle_{x\sim P(x)} + \langle g(y) \rangle_{y\sim Q(y)} - T \int dxdy\, \pi^*(x, y) \Big). \tag{6.83}$$

---

[13] This generalization is not necessary to derive the final form of the WGAN, but considering the Helmholtz free energy makes the derivation easier to understand.

[14] The term "+1" is not necessary but it will make the later discussion cleaner.

[15] For example, there is a way to reduce the amount of calculation in the actual calculation algorithm compared to the original zero temperature problem [85].

Such a situation that the original problem (min) becomes equivalent to some other problem (max) often appears in machine learning,[16] and it is called a **strong duality** in optimization problems. Looking at the derivation here, we see that the new degree of freedom of the dual problem is the Lagrange multiplier that expresses the constraint, and it is similar to the duality in statistical mechanics and field theories. At the time of writing this, the authors cannot tell if this is just a similarity or has a profound meaning.

Kantorovich-Rubinstein duality

Consider taking a zero temperature limit $T \to +0$ to return to the original problem (6.76). The dual problem appears to be able to reach the limit without any difficulty. Solving (6.82) gives

$$\pi^*(x, y) = e^{\frac{-H(x,y)}{T}}, \tag{6.84}$$

$$H(x, y) = E(x, y) - f(x) - g(y). \tag{6.85}$$

Since (6.83) is about the maximum value, from the beginning, it is better to remove f, g with which the last term becomes $-\infty$. According to this argument, we find a condition similar to (6.51), where the "Hamiltonian" (6.85) is bounded from below,

$$H(x, y) \geq 0, \quad \text{i.e.,} \quad f(x) + g(y) \leq E(x, y). \tag{6.86}$$

Therefore,

$$D_W(P, Q) = \max_{f(x)+g(y)\leq E(x,y)} \Big(\langle f(x) \rangle_{x\sim P(x)} + \langle g(y) \rangle_{y\sim Q(y)} \Big). \tag{6.87}$$

In particular, if the energy function for transport is taken as $E(x, y) = ||x - y||$, at least $g = -f$ must be attained to achieve the maximum, so finally

$$D_W(P, Q) = \max_{f(x)-f(y)\leq ||x-y||} \Big(\langle f(x) \rangle_{x\sim P(x)} - \langle f(y) \rangle_{y\sim Q(y)} \Big). \tag{6.88}$$

In other words, the point is that f is restricted to functions with Lipschitz continuity.

WGAN

Incorporating idea of GAN into this duality leads to the idea of Wasserstein GAN. Simply we set $Q = Q_G$ and optimize G to minimize the Wasserstein distance:

$$\min_G D_W(P, Q_G) = \min_G \max_{f(x)-f(y)\leq ||x-y||} \Big(\langle f(x) \rangle_{x\sim P(x)} - \langle f(y) \rangle_{y\sim Q(y)} \Big). \tag{6.89}$$

[16] As another example, support vector machines (which are not described in this book) have also a duality.

Then, adopting $f = D$, we find that this is a minimax problem (6.41) in the zero-sum GAN! Namely, for

$$V_D(G, D) = \langle D(x) \rangle_{x \sim P(x)} - \langle D(y) \rangle_{y \sim Q(y)}, \tag{6.90}$$

$$V_G(G, D) = -V_D(G, D), \tag{6.91}$$

the equilibrium point of the GAN is also expected from the property of $D_W(P, Q_G)$ to satisfy

$$P(x) = Q_G(x). \tag{6.92}$$

One should note that here D must be Lipschitz-continuous. When making D with a neural network, normally this condition is not satisfied. The original paper implements an approximate Lipschitz continuity by restricting the range of the values of the weights. And later years there appeared some ideas such as an approximate implementation with a gradient penalty $(||\nabla_x D(x)||_2 - 1)^2$ for D as a regularization term [86], or a spectral normalization in which the network weights are normalized by their maximum singular value [87].[17] The gradient penalty and the spectral normalization are also known to improve the performance of other GANs which are not WGAN.

6.4 Generalization in Generative Models

So far, we have introduced various models and algorithms of unsupervised machine learning and deep learning.[18] All of those are aimed to make

$$Q_{J^*}(x) = P(x). \tag{6.93}$$

However, as we have emphasized many times, we cannot access the data generation probability distribution $P(x)$ in machine learning in the first place. So the learning algorithm has to use the empirical distribution $\hat{P}(x)$ for $P(x)$, and how should we think about generalization?

[17] Actually, spectral normalization was introduced to stabilize the learning process of conventional GANs rather than to use it for WGANs, and the authors are not aware of successful examples of using spectral normalization in WGAN implementations. With this normalization, the network acquires a K-Lipschitz continuity with a certain number K, so there is no reason why it cannot be used for WGAN.

[18] Noteworthy deep generative models that we have not been able to introduce here include variational auto-encoder (VAE) [88, 89] and nonlinear independent component estimation (NICE) [90]. A brief review is provided by a physicist, L. Wang [91].

Case of rote memorization
First, if we simply replace (6.93) with the empirical distribution,

$$Q_{J^*}(x) = \hat{P}(x)\,, \tag{6.94}$$

it simply appears that "the machine remembers the data that came out." Actually, we want the machine to create new images that do not exist in the data, so this is not sufficient. However, there are things we can learn from this setup. We have already derived an inequality of generalization performance in unsupervised machine learning of the rote memorization: it is (5.21) which was explained at the end of the section for the central limit theorem. Here we repeat explaining the problem again. An event with an occurrence probability of $p_1, p_2, \ldots, p_W$ is defined as $A_1, A_2, \ldots, A_W$, and the specific value of the probability p_i is not known. Instead, we have only the data

$$\bullet \begin{cases} \text{The number of event } A_1 \text{ is } \#_1, \text{ the number of } A_2 \text{ is } \#_2, \ldots, \text{ the number of } A_W \text{ is } \#_W, \\ \text{In total, the number of events is } \# = \sum_{i=1}^{W} \#_i\,. \end{cases} \tag{6.95}$$

and the problem is to estimate the value of p_i. This problem can be thought of as an unsupervised learning. Obviously, the appropriate estimated probability is the "empirical distribution" $q_i = \frac{\#_i}{\#}$, and if this probability is close to p_i, it can be said that the model is generalized. The central limit theorem says

$$\begin{aligned} &\text{With approximately 70\% probability,} \\ &\text{one finds } p_i - \sqrt{\frac{p_i(1-p_i)}{\#}} < q_i < p_i + \sqrt{\frac{p_i(1-p_i)}{\#}}\,. \end{aligned} \tag{6.96}$$

Therefore, even in the case of the unsupervised learning, the generalization performance is expected to be of the order of $1/\sqrt{\#}$.

Actual cases
Do actual deep generative models memorize everything by rote? In the case of using the actual empirical distribution $\hat{P}(x)$, the proof for (6.93) of GANs can be applied exactly as it is, and it appears that (6.94) is proven. However, there are other differences between the proof and the actual cases:

- The convergence proof of GANs assumes that G and D have infinite expressive power, while in the actual training, G and D are deep neural networks with a fixed structure, and their expression is limited.
- The Nash equilibrium condition (6.39), (6.40) which is assumed in the proof is actually solved by the gradient method (6.37), (6.38). However, the gradient method is not exact, and therefore, the obtained trained model is not an exact equilibrium point.

Fig. 6.1 Left: Images created by G. Right: Image created by G (upper left red frame) and 9 images closest to it in the training data

Therefore, it is too early to say that the proof of (6.93) under ideal assumptions is actually "only" (6.94), which is a replacement with an empirical distribution. Then, what is the conclusion, in the end? A picture is worth a thousand words. Here, we show the image generated by a GAN which was actually trained with CIFAR-10, in Fig. 6.1 (left). With this alone, we cannot tell more than "something that looks like a natural image is generated," so in Fig. 6.1 (right) we show the image generated by the GAN and the training data in the neighborhood of it found in the image space. A glance finds no image which looks exactly the same as the image generated by G, and furthermore, even in the closest image the detailed structure is different. This is supporting evidence that the GAN, a kind of deep generative model, avoids a rote memorization (6.94) and generalizes.

Inception score

Let us introduce an attempt to measure a "generalization performance" in image generation from a completely different perspective. In the first place, we regard (6.93) as a state which generalizes, because the sampling from this generator gives us a "realistic" image x. Hence the generalization performance is whether images that "look real to humans" can be generated or not. On the other hand, many "image classification networks" led by ResNet [24] trained using the ImageNet dataset [26] are now known to have classification accuracy superior to that of humans [92]. So the idea is to have the image classification network decide whether the generated image is "real-looking to humans" or not. Let us prepare

$$\text{Probability distribution given by generator: } Q_{G^*}(x), \tag{6.97}$$

$$\text{Trained image classification network: } Q_{J^*}(d|x). \tag{6.98}$$

What quantity should be considered using these two? First, suppose the generator draws some image:

$$x_{\text{fake}} \sim Q_{G^*}(x). \tag{6.99}$$

Using this image, what is the probability $Q_{J^*}(d|x_{\text{fake}})$ of the classification label d? For simplicity, let d = (dog, cat). As a first example, suppose the generated image is completely useless, and x_{fake} looks like a mysterious animal between dogs and cats. Then the image looks like a cat and like a dog, so

$$Q_{J^*}(d = \text{dog}|x_{\text{fake}}) \approx \frac{1}{2}, \quad Q_{J^*}(d = \text{cat}|x_{\text{fake}}) \approx \frac{1}{2}. \tag{6.100}$$

On the other hand, if x_{fake} is an image which looks really like a dog, we should have

$$Q_{J^*}(d = \text{dog}|x_{\text{fake}}) \approx 1, \quad Q_{J^*}(d = \text{cat}|x_{\text{fake}}) \approx 0. \tag{6.101}$$

The entropy in each case is

$$\begin{aligned} S(x_{\text{fake}}) &= -\sum_d Q_{J^*}(d|x_{\text{fake}}) \log Q_{J^*}(d|x_{\text{fake}}) \\ &\approx \begin{cases} \log 2 & (x_{\text{fake}} \text{ is a bad image (6.100)}), \\ 0 & (x_{\text{fake}} \text{ is a good image (6.101)}). \end{cases} \end{aligned} \tag{6.102}$$

Namely, the closer $S(x_{\text{fake}})$ is to 0, the more reality it has. So, this quantity is the appropriate one to calculate.

But here is a trap. Equation (6.102) certainly measures the reality, but it only measures goodness for a single x_{fake}. For example, by sampling from the generator, we obtain many images

$$x_{\text{fake1}}, x_{\text{fake2}}, \ldots, \sim Q_{G^*}(x), \tag{6.103}$$

and if most of the images are similar dog images, the individual (6.102) values are certainly large, but it is not possible to create a new image, so, resultantly, the machine has not generalized. In other words, generators are required to generate realistic images that are "sufficiently diverse." Entropy can also be used to measure such diversity. We just need to calculate the entropy of the expectation value of the classification probability when various x_{fake} are input,

$$Q(d) = \int dx \, Q_{J^*}(d|x) Q_{G^*}(x) = \langle Q_{J^*}(d|x) \rangle_{x \sim Q_{G^*}(x)}. \tag{6.104}$$

Explaining again with the example of dogs and cats, if $Q_{G^*}(x)$ generates both dog and cat images equally, the probability for x_{fake} to be classified as dog or cat is $\frac{1}{2}$,

$$Q(d=\text{dog}) \approx \frac{1}{2}, \quad Q(d=\text{cat})\,. \approx \frac{1}{2}. \tag{6.105}$$

On the other hand, if $Q_{G^*}(x)$ generates only dogs as in the case above, we have

$$Q(d=\text{dog}) \approx 1, \quad Q(d=\text{cat}) \approx 0\,. \tag{6.106}$$

Then the entropy of $Q(d)$ is

$$\begin{aligned} S &= -\sum_d Q(d) \log Q(d) \\ &\approx \begin{cases} \log 2 & (Q_{G^*} \text{ is diverse, (6.105)}), \\ 0 & (Q_{G^*} \text{ is not diverse, (6.106)}). \end{cases} \end{aligned} \tag{6.107}$$

We see that the larger S, the more diverse.

Taken together, a good generator should decrease (6.102) and increase (6.107). Since entropy is additive, the "score" of the generator can be thought of as the difference. Then this can be written as an expectation value of a relative entropy.[19]

$$\begin{aligned} & S - \langle S(x) \rangle_{x \sim Q_{G^*}(x)} \\ &= \sum_d \Big(\underbrace{\langle Q_{J^*}(d|x) \log Q_{J^*}(d|x) \rangle_{x \sim Q_{G^*}(x)}}_{\text{Transform this into an integration}} - \underbrace{Q(d)}_{(6.104)} \log Q(d) \Big) \\ &= \sum_d \Big(\int dx\; Q_{G^*}(x) Q_{J^*}(d|x) \log Q_{J^*}(d|x) - \int dx\; Q_{G^*}(x) Q_{J^*}(d|x) \log Q(d) \Big) \\ &= \int dx\; Q_{G^*}(x) \Big(\sum_d Q_{J^*}(d|x) \log \frac{Q_{J^*}(d|x)}{Q(d)} \Big) \\ &= \int dx\; Q_{G^*}(x) D_{KL}\Big(Q_{J^*}(d|x) \Big|\Big| Q(d) \Big) \\ &= \Big\langle D_{KL}\Big(Q_{J^*}(d|x) \Big|\Big| Q(d) \Big) \Big\rangle_{x \sim Q_{G^*}(x)}. \end{aligned} \tag{6.109}$$

[19] Yet another transform

$$(6.109) = \int dx \sum_d Q_{J^*G^*}(x,d) \log \frac{Q_{J^*G^*}(x,d)}{Q_{G^*}(x) Q(d)}, \quad Q_{J^*G^*}(x,d) = Q_{J^*}(d|x) Q_{G^*}(x) \tag{6.108}$$

provides a quantity called mutual information between the generated image and the classification labels. Reference [93] used this to hack IS, that is, generate images that have an unusually large value of the IS (although images that make little sense to the human eyes). As seen from this example, it is not always true that the higher the IS, the better.

Fig. 6.2 $IS(\text{Left}) = 8.54$, and $IS(\text{Right}) = 36.8$ (official value)

As in the Sanov theorem introduced in Chap. 1, the exponentiation of this quantity is called the inception score, the score of the generator Q_{G^*} in image generation [94]:

$$IS(Q_{G^*}) = \exp\Big\langle D_{KL}\Big(Q_{J^*}(d|x)\Big|\Big|Q(d)\Big)\Big\rangle_{x\sim Q_{G^*}(x)}. \tag{6.110}$$

This score is calculated using a sampling approximation from $Q_{G^*}(x)$, based on the law of large numbers. It is one of the de facto standards for measuring GAN metrics.[20] This "Inception" is a neural network module in the structure of GoogLeNet [96], a winning model by Google in the ImageNet classification competition in 2014,[21] and this GoogLeNet is used exclusively for the calculation of IS.

Figure 6.2 shows the generated images of a GAN trained with STL-10 [97] and of the model trained with ImageNet, which is freely available [87, 98] (https://github.com/pfnet-research/sngan_projection). We can see that the IS is high when the generator can generate images well.

[20]There is another index called Fréchet inception distance (FID) [95], which is the Wasserstein distance (which is called the Fréchet distance, and the name FID follows from it) between the data image distribution and the generated image distribution in the hidden layer (feature space) of image classification network, assuming that the features in a classification network follow a certain Gaussian distribution.

[21]The name Inception came from the title of a popular Hollywood movie "Inception": the title of the GoogLeNet paper is "Going deeper with convolutions," while the main character's line in the movie is "We need to go deeper." The original paper even cites the movie (the article summarizing the backgrounds). It is a witty naming that makes anyone who has seen this movie grin.

The highest value of IS for ImageNet as of March 2019 is probably $IS = 166.3$ achieved in Ref. [99]. Interested readers should take a look at the image given in the first page of the paper (which can be read for free). We are sure that the readers will be surprised at how exquisite it is. For reference, if one uses $Q_{J^*}(d|x)$ which processes a task classifying 1000 labels (classification using ImageNet), the full mark for the score is $IS = 1000$ [93, 95].

Column: Self-Learning Monte Carlo Method

According to Ref. [100], the contrastive divergence method is a kind of optimization of an "error function"

$$K_{ex}(\theta) = D_{KL}\Big(P_\theta(\mathbf{x}'|\mathbf{x})P(\mathbf{x}) \Big|\Big| P_\theta(\mathbf{x}|\mathbf{x}')P(\mathbf{x}') \Big) . \tag{6.111}$$

At $\theta = \theta_0$ where this quantity vanishes, the detailed balance condition is satisfied, so the target distribution $P(\mathbf{x})$ is a convergence destination of a Markov chain $P_{\theta_0}(\mathbf{x}'|\mathbf{x})$. A little massage of this equation using Bayes' theorem gives

$$\begin{aligned} K_{ex}(\theta) &= D_{KL}\Big(P_\theta(\mathbf{x}'|\mathbf{x})P(\mathbf{x}) \Big|\Big| P_\theta(\mathbf{x}'|\mathbf{x}) \frac{e^{-H_\theta^{\text{eff}}(\mathbf{x})}}{e^{-H_\theta^{\text{eff}}(\mathbf{x}')}} P(\mathbf{x}') \Big) \\ &= -\Big\langle \log \frac{P(\mathbf{x}')e^{-H_\theta^{\text{eff}}(\mathbf{x})}}{P(\mathbf{x})e^{-H_\theta^{\text{eff}}(\mathbf{x}')}} \Big\rangle_{P_\theta(\mathbf{x}'|\mathbf{x})P(\mathbf{x})} \geq 0 . \end{aligned} \tag{6.112}$$

The last inequality is due to the property of relative entropy. Therefore, roughly speaking, bringing the following quantity closer to 1,

$$\frac{P(\mathbf{x}')e^{-H_\theta^{\text{eff}}(\mathbf{x})}}{P(\mathbf{x})e^{-H_\theta^{\text{eff}}(\mathbf{x}')}} \tag{6.113}$$

is the contrastive divergence method. As a matter of fact, the transition probability obtained by the heatbath method together with the Metropolis test using this factor satisfies exactly the detailed balance condition.

$$\begin{aligned} P_\theta^{ex}(\mathbf{x}'|\mathbf{x}) &= \min\Big(1, \frac{P(\mathbf{x}')e^{-H_\theta^{\text{eff}}(\mathbf{x})}}{P(\mathbf{x})e^{-H_\theta^{\text{eff}}(\mathbf{x}')}}\Big) P_\theta(\mathbf{x}'|\mathbf{x}) \\ &= \frac{P(\mathbf{x}')e^{-H_\theta^{\text{eff}}(\mathbf{x})}}{P(\mathbf{x})e^{-H_\theta^{\text{eff}}(\mathbf{x}')}} \min\Big(\frac{P(\mathbf{x})e^{-H_\theta^{\text{eff}}(\mathbf{x}')}}{P(\mathbf{x}')e^{-H_\theta^{\text{eff}}(\mathbf{x})}}, 1\Big) \frac{e^{-H_\theta^{\text{eff}}(\mathbf{x}')}}{e^{-H_\theta^{\text{eff}}(\mathbf{x})}} P_\theta(\mathbf{x}|\mathbf{x}') \\ &= \frac{P(\mathbf{x}')}{P(\mathbf{x})} P_\theta^{ex}(\mathbf{x}|\mathbf{x}') . \end{aligned} \tag{6.114}$$

In this way, the desired convergence is guaranteed even if the probability defined by H_θ^{eff} does not exactly match the target P. Note that, to implement this correction term, one needs to have access to the value of $P(\mathbf{x})$. For example, if the target is written by some Hamiltonian

$$P(\mathbf{x}) = \frac{e^{-H_{\text{true}}(\mathbf{x})}}{Z}, \tag{6.115}$$

it is possible. In this case, the training sample is a snapshot of the configuration from the statistical mechanical system defined by this Hamiltonian. A machine learning Metropolis method that repeats the cycle of (1) training H_θ^{eff} with the configurations generated by Markov chain Monte Carlo method for (6.115), and (2) generating new configurations with a Markov chain of type (6.114), is called a self-learning Monte Carlo method, which has been actively studied since 2016 [101].

Part II
Applications to Physics

The second part describes the application of machine learning to theoretical physics, which has recently begun, with examples and history. Because machine learning is one of the new techniques in science, its involvement in physics is diverse. From the standard viewpoints in physics: "inverse problems", "phases", "differential equations", "quantum many systems", and "spacetime", we will look at possible standpoints of machine learning, historical development of it, and some of the recent developments. This understanding of machine learning from the perspective of physics will provide readers' perspectives on the relationship between machine learning and physics so far and in the future, and will be helpful for study and research.

Chapter 7: Inverse Problems in Physics First, we consider inverse problems in physics. In fact, inverse problems are at the heart of revolutionary development in physics. What does it mean to solve an inverse problem? What is the meaning of the phrase "machine learning is good at solving inverse problems"? You will gain a comprehensive perspective and significance in applying machine learning to theoretical physics.

Chapter 8: Detection of Phase Transition by Machines As an approach to the important question of whether machines can learn the discovery of physics, this chapter examines the question "Can phase transitions be found by deep learning?" Understanding phases is one of the most important subjects in physics. Can machine learning really discover the thermal phase transition in the basic physical system: the Ising model?

Chapter 9: Dynamical Systems and Neural Networks Neural networks are a way of expressing a variety of nonlinear functions, but can also be thought of as waves of information propagating between layers. In this chapter, we show that such multi-layer propagation can be interpreted as the time evolution of dynamical systems, and hence of Hamiltonian systems, and look at the close relationship between the fundamental concept of "time evolution" in physics and deep neural networks.

Chapter 10: Spinglass and Neural Networks To begin with, neural networks are based on neural circuits formed by neurons in human brains. One of the important mechanisms of the brain is memory. The Hopfield model, which explains the mechanism of memory in terms of physics, is a bridge between physics and neural networks. In this chapter, we explain the Hopfield model and investigate the relationship between machine learning and spin glass, which is still a rich subject in condensed matter physics.

Chapter 11: Quantum Manybody Systems, Tensor Networks and Neural Networks In condensed matter physics, finding the wave function of a quantum many-body system is the most important issue. Theoretical development in recent years includes a wave function approximation using a tensor network. At first glance, it is very similar to neural network diagrams, and how are they actually related? In this chapter we will see the relation and mapping, and that the restricted Boltzmann machine is closely related to tensor networks.

Chapter 12: Application to Superstring Theory The last chapter describes an example of solving the inverse problem of string theory as an application of deep learning. The superstring theory unifies gravity and other forces, and in recent years, the "holographic principle," that the world governed by gravity is equivalent to the world of other forces, has been actively studied. We will solve the inverse problem of the emergence of the gravitational world by applying the correspondence to the dynamical system seen in Chap. 9, and look at the new relationship between machine learning and spacetime.

These chapters, from Chaps. 7 to 12, can be read almost independently. You can pick up a section that interests you. Reading all the chapters will lead the readers to find a clear understanding of the relationship between machine learning and physics through its various examples and uses in physics, and through its history.

Chapter 7
Inverse Problems in Physics

Abstract First, we consider inverse problems in physics. In fact, inverse problems are at the heart of revolutionary development in physics. What does it mean to solve an inverse problem? What is the meaning of the phrase "machine learning is good at solving inverse problems"? You will gain a comprehensive perspective and significance in applying machine learning to theoretical physics.

There are many **inverse problems** in the world, and so there are in physics. In this chapter, we will look at inverse problems[1] that can appear in physics from the perspective of the machine learning, and learn the significance of them.[2] What is an "inverse problem," in the first place? In general, machine learning is often said to be good at solving inverse problems. Why is that? By examining these, we can concretely see how machine learning and deep learning can be applied to physics, and at the same time we will learn how it makes sense to rely on machine learning for individual physics problems that individual researchers have.

7.1 Inverse Problems and Learning

The inverse of the usual approach to solving a problem is commonly called the inverse problem. The general nature of the inverse problem will be described later, and let us start with a simple example.

Consider a classical mechanical problem that follows the evolution of time. Given a differential equation that determines the time evolution, and then given the initial state at a certain time $t = 0$, the state at any time $t > 0$ is determined. The

[1]The inverse problem in mathematics is different from the generic inverse problem dealt with here. In mathematics, when there is a theorem C that A is B, then the proposition that B is A is called the inverse of theorem C. For example, one of the famous theorems proved by Kiyoshi Oka is Hartogs' inverse problem, which is the inverse problem in that sense.

[2]The contents of this chapter do not enumerate general solutions for inverse problems.

A. Tanaka et al., *Deep Learning and Physics*, Mathematical Physics Studies,
https://doi.org/10.1007/978-981-33-6108-9_7

Schrödinger equation in quantum mechanics is considered the same way. On the other hand, as its "inverse," we can ask the following question: given the state at a certain time $t > 0$, is it possiblc to determine the state at the time $t = 0$ to reproduce it? This is what is called the initial value problem, and it is an inverse problem.

This example of a time evolution is easy to understand. More generally, the inverse problem can be formulated as follows. Consider a polynomial function with real number x as an argument:

$$f(x) = \sum_{k=0}^{n} J_k x^k . \tag{7.1}$$

Given x, $f(x)$ is determined, so the procedure to calculate $f(x)$ from x can be thought of as "time evolution" in which, given an initial value of x, after a while, the system gives $f(x)$. Thus, the inverse problem of "going back in time"İ is that, when a function $f(x)$ is given and a value of $y = f(a)$ is given, "solve a" is the problem. Causally, it is a matter of estimating the cause from the result.[3]

When considering the temporal order, there is a causal relationship between the cause and the result. There are many inverse problems that are not about the temporal order, but about spatial order. For example, nondestructive inspection and X-ray tomography are often said to be typical examples of inverse problems. This is a method of estimating whether there is a cavity inside an object by using laws of conduction. Given the external shape of the object (outer surface boundary condition, $y = f(a)$ in the above equation) and the laws of conduction (e.g. Poisson equation, wave equation, heat conduction equation, etc. which give the form of the function $f(x)$, that is, the coefficient J_k), the problem is to find the shape of the internal cavity, that is, x. In this way, the problem of finding the original initial conditions or boundary conditions from the values, for a given system of equations, is an inverse problem.

Now, there is another kind of inverse problem, namely, when the function $f(x)$ itself is unknown, find the function $f(x)$. In this case, the given condition is that when the input data is $x = x_1$, the output data is $y_1 = f(x_1)$. Given such a condition, the problem is to determine the unknown function $f(x)$.

Of course, just a condition $y_1 = f(x_1)$ cannot determine a general function. This pair (x_1, y_1) only has one constraint expression for the unknown coefficients $J_0, \cdots, J_n$ of the polynomial function. Therefore, to determine the unknown function $f(x)$, generally the same number of input/output pairs (x_i, y_i) $(i = 1, 2, 3, \cdots)$ must be prepared.

Supervised machine learning is this latter problem. The pair of input data and output data becomes the training data, and the unknown function becomes a neural

[3]Of course, the Schrödinger equation is a linear equation, so positive time evolution and negative time evolution are the same in terms of difficulty to solve, and should not be called inverse problems.

network. In other words, the coefficients of the unknown function correspond to the weights of the neural network.

If the form of the function $f(x)$ is known to a certain extent, the term "modeling" is often used, rather than machine learning. Due to physical requirements, the shape of $f(x)$ is restricted. For example, if we know the basic behavior such as a heat diffusion system and wave system, and if external field fluctuations and other interactions are added to it, the basic form of $f(x)$ is determined, and one will write the model in a perturbative way with small effects. The additional effect adds a new term to the function, so its coefficient is unknown. This unknown coefficient is determined from the input data and the output data. For example, if it is known that $f(x)$ is a linear function of x, a linear regression suffices.

On the other hand, in the case of deep learning, since the function form is versatile,[4] we deal with a very wide class of functions without specifying the model. And by the training of the machine, we gradually limit the model, that is, the shape of the function. Therefore, machine learning and deep learning are the inverse problems of the latter type among the above two meanings, and can be said to be those which do not assume a model (function form) as its unknown function.

In this way, the nature of the inverse problem in the map (7.1) greatly depends on which part is considered unknown. Learning is considered to be a kind of inverse problems as described above. But when applying it to a physical system or exploring an analogy with a physical system, one needs to be clear about what one is trying to solve and which part is unknown.

The general properties of a problem called the "inverse problem" include the following:

- Knowing objects that cannot be measured directly
- Infering the cause from the results
- Determining physical laws and governing equations
- Determining physical constants

An inverse problem is one of the above, requiring a way to solve it in the reverse direction. Of these, things like "determining the laws of physics" are the most important things in physics, and it is no exaggeration to say that all physics innovations are inverse problems. Johannes Kepler discovered his third law of planetary motion from Tycho Brahe's precise planetary observation data. Max Planck discovered the radiation formula. The underlying regularity in the data sublimated to the law, which is the inverse problem. The importance of it is obvious.

Machine learning is a technique that has the potential to apply to all four of these properties. Of course, the underlying mechanism, such as generalization, is not well understood yet, but instead of relying solely on the physical sense of those called "genius physicists," machine learning provides a powerful and general method for inverse problems.

[4] Refer to Chap. 3 for the universal approximation theorem.

7.2 Regularization in Inverse Problems

Now, to look more closely at the relationship between machine learning and inverse problems, consider the case of a linear function, instead of the nonlinear function (7.1), where the nature of the inverse problem is clearer. As we will see below, there are ways to solve inverse problems well for linear function problems. Along the way, we introduce the important concept called **regularization**.

First, consider the following linear transformation:

$$y_i = \sum_{k=1}^{m} J_{ik} x_k \,. \tag{7.2}$$

Here, the input or the initial value is the vector x_k. Let its dimension be m. If the dimension of the output y_i is n, J is a matrix of the size $n \times m$.

In this linear equation (7.2), consider the usual (the first one in the above sense) inverse problem of finding x_k for a given y_i. In the case of a square matrix $n = m$, the story is simple. If the inverse matrix J^{-1} of the matrix J is obtained, then x_k is obtained as follows:

$$x_k = \sum_{i=1}^{m} \left[J^{-1} \right]_{ki} y_i \,. \tag{7.3}$$

While this inverse problem seems to work, there are actually two difficulties that can arise:

- First, when the determinant of the matrix J is very close to zero numerically, the difficulty is that x_k changes significantly even for slightly different values of y_i.
- Second, if the number of data y_i is not enough, it is difficult to determine x_k even if there is an inverse matrix. This can be considered equivalent to the case $n < m$.

These issues are related in terms of data errors. When data is actually handled, it always comes with fluctuations and errors of the measurements. Taking these factors into account, solving the inverse problem can have such difficulties, even for a simple linear problem.

Generally, **well-posed problems** defined by J. S. Hadamard refers to a problem that satisfies the following three properties:

(1) Existence of a solution
(2) Uniqueness of the solution
(3) Stability of the solution (the solution changes infinitesimally when the initial condition changes infinitesimally)

Problems where any of these are not met are called **ill-posed problems**. The inverse problem which we described as the first example in the above is an ill-posed problem.

In many cases, although the forward problem satisfies the well-posedness, the inverse problem is ill-posed. A typical example is the process of the many-body system evolving to a macroscopic equilibrium. In a normal physical system in which relaxation to an equilibrium state occurs, such as a thermal diffusion system, in order to obtain the initial state before time evolution from the state after time evolution, a very small fluctuation of the output value (from a perfect equilibrium state) must be measured. In other words, relaxed systems generally do not have the stability (3).

How can we deal with ill-posed problems? For example, when $n > m$ in the equation system (7.2), it is an **overdetermined system**, and solutions cannot be determined in that form. To obtain a "close solution" x_i in such a situation, consider the following. First, rewrite (7.2) to the equivalent equation:

$$L \equiv \sum_{i=1}^{n} \left(y_i - \sum_{k=1}^{m} J_{ik} x_k \right)^2 = 0 \,. \tag{7.4}$$

There is generally no solution for this equation, but instead of finding a solution, we seek an x_k that minimizes L. This is exactly the least squares method.

Tikhonov's regularization method is a general method to treat ill-posed problems. Let's consider the case where the number of conditions is not enough ($n < m$), instead of the overdetermined system. There are countless solutions x_k that satisfy Eq. (7.2), but a solution is required that has the properties one wants. This "property" depends on the physical background of the problem. For example, one may require a physical property that x_k's should better to have the same order of magnitude, or that there should not be much difference between them. To add the condition that any of the x_k's should not be much bigger than the other, add a regularization term to the L and find the x_k so that the following L' is minimized:

$$L' \equiv L + \alpha \sum_{k=1}^{m} (x_k)^2 \,. \tag{7.5}$$

Then, from the myriad of solutions, one can get the solution with the properties one wants.[5] In an ill-posed problem, adding a regularization term and changing the problem to a well-posed one in this way is called Tikhonov's regularization method. The regularization term to be added may be the L2 norm as described above or the L1 norm.[6] If a condition of no rattling is necessary, the following regularization

[5]The coefficient α is called a regularization parameter, and from the learning point of view it is called a hyperparameter. Choosing hyperparameters generally depends on what solution one wants and how to make learning more efficient, and it also prevents machines from over-training. A method to determine the hyperparameter α from the guideline that the nature of the solution (for example, the size of $|x|$ in the case of (7.4)) is about the same as the size of the input/output error, is called **Morozov's discrepancy principle**.

[6]These cases are called sparse modeling, and applied to physical systems [102].

term may be added:

$$\Delta L \equiv \alpha \sum_{k=1}^{m-1} (x_k - x_{k+1})^2 \,. \tag{7.6}$$

One way to find a solution that minimizes (7.5) is the singular value decomposition (SVD). Decompose the matrix J as follows:

$$J_{ik} = \sum_{p=1}^{P} \mu_p u_i^{(p)} v_k^{(p)} \,. \tag{7.7}$$

Here the vector u and v are orthogonal bases satisfying

$$\sum_{i=1}^{n} u_i^{(p)} u_i^{(q)} = \delta_{p,q} \,, \quad \sum_{i=1}^{m} v_k^{(p)} v_k^{(q)} = \delta_{p,q} \,. \tag{7.8}$$

P can be taken to satisfy $n \geq P$, $m \geq P$. Such (7.7) is called the singular value decomposition of J. Using this singular value decomposition, the solution x_k that minimizes (7.5) is known to be constructed as follows:

$$x_k = \sum_{i=1}^{n} A(\alpha)_{ki} y_i \,, \tag{7.9}$$

$$A(\alpha)_{ki} \equiv \sum_{p=1}^{P} \frac{\mu_p}{(\mu_p)^2 + \alpha} u_i^{(p)} v_k^{(p)} \,. \tag{7.10}$$

The matrix $A(\alpha)$ is a generalization of the inverse matrix, and is called the **Moore–Penrose inverse matrix**, especially when $\alpha = 0$.

Thus, in the case of linear transformation, even if the inverse problem is ill-posed, there is a general solution using regularization. However, in the case of machine learning, the number of parameters and the amount of data are enormous, so it is a large-scale underdetermined system or a overdetermined system, and since the function is nonlinear, there is no analytic solution. Therefore, one has to search for a solution by numerical calculation that incorporates various regularization methods.

7.3 Inverse Problems and Physical Machine Learning

As described so far, there are two types of inverse problems: boundary value problems and system decision problems. All of these are issues that frequently appear in physics and are important scenes in actual research.

The boundary value problem does not work well when there is a stability problem, among the ill-posedness according to Hadamard. As one of such examples, we described the thermal diffusion equation. More generally, the inverse problem of chaotic systems has stability problems. A chaotic system is a deterministic dynamical system with bounded orbits and a small difference in initial conditions shows an exponential increase. In other words, it is a system that "gets messed up over time and does not remember the initial state." Given the final state, one will not be able to go back in time to build the initial state.

Nevertheless, the research of **chaos** is progressing in various ways. The reason is that there are research perspectives regarding what kind of decision equation of time evolution can produce chaos, and what is the source. For example, there is a proof that chaos does not occur if the number of degrees of freedom is too small: the Poincaré–Bendixson theorem. Therefore, even if the boundary value inverse problem is ill-posed, the target of research could be fertile, putting the actual problem of solvability aside.

For example, there are many studies that show that chaotic attractors are related to the mechanism of memory in the brain, so research related to neural networks can be expected in the future. In addition, the **Lyapunov exponent**, which characterizes the strength of chaos (the Lyapunov exponent λ measures the extent to which the difference exponentially amplifies as $e^{\lambda t}$), gives a quantitative idea of how ill-posed the inverse problem is. The Lyapunov exponent of quantum systems is being studied in relation to the AdS/CFT correspondence in string theory and quantum information theory, and further development about the relationship with machine learning is expected.

On the other hand, the inverse problem which is categorized as the **system determination problem** is exactly where machine learning is applied. A system determination problem that has been studied for a long time is a potential determination problem in quantum mechanics. Here we describe one example of the inverse problem in quantum mechanics.

The quantum mechanics of a one-body system is defined by a Hamiltonian H. Given the initial state wave function, the final state wave function can be obtained by the time evolution operator e^{-iHt} with the Hamiltonian H. This is a forward problem. The inverse problem as a system determination problem is the case when the Hamiltonian H is unknown. What we should regard as known is important to solve this problem. As an example, let us describe one of the well-studied methods, the inverse scattering method. In the inverse scattering method, when the unknown part of the Hamiltonian is the potential $V(x)$ and the potential is asymptotically flat, the potential is determined from information such as scattering amplitudes. Suppose a one-dimensional non-relativistic quantum mechanics is given by an unknown potential $V(x)$. Suppose also that we measured the S matrix that tells us how much of a plane wave coming from the spatial infinity is transmitted and reflected. In addition, suppose we know how the wave function bound to potential $V(x)$ approaches zero near infinity. A method for reconstructing the potential $V(x)$ from this information is the inverse scattering method. In the inverse scattering method, the potential function $V(x)$ can be constructed by solving a certain integral equation from these data.[7]

In the inverse scattering method, an unknown potential, or Hamiltonian, was constructed from asymptotic data on scattering and bound states. Then, what about a system that has only bound states? In this case, the Hamiltonian eigenvalues will be the known data. Can we reconstruct the Hamiltonian from energy eigenvalues? There has been a wide variety of such motivated studies for many years. For example, when exploring superconducting materials, the interaction Hamiltonian that can produce a superconducting state even at high temperatures is being sought. In a broad sense, drug discovery is the same type of problem. As can be seen from these motivations, we need to model the Hamiltonian to some extent and determine its coefficients from the data. Recent advances in machine learning research have shown that machine learning has great potential for determining models and determining coefficients.

The holographic principle in string theory (AdS/CFT correspondence), which will be described as an example in Chap. 12, is also an inverse problem as a system determination problem. It can also be a boundary value problem in the sense of determining gravity theory, the inside (bulk), from the quantum field theory "living" on the boundary. In the following chapters, we will also introduce how to view the neural network itself as the time axis of the time evolution of dynamical systems, and how to view it as an emergent space in the AdS/CFT correspondence.

A closer look at the relationship between deep learning and physics may reveal solutions to the "bottleneck" of inverse problems in various physics problems. Formulating the inverse problem is essentially the way of discovering the fundamentals of science. So it is natural to expect that the discovery of new laws will come from revealing the relationship between machine learning and physics.

Column: Sparse Modeling

In this column, let us take a look at a technique called **sparse modeling**. Sparse modeling is a research field of image sensing such as the one for medical purposes[8] and has been used for nearly 20 years. Some readers may have heard that recently it was used for the "method that captures the shadow of a black hole" [103]. By using this method, one can extract beautiful images from noisy data.[9]

We take a closer look at Tikhonov's regularization method described in this chapter. Generally, an equation system in which the number of unknowns is larger than the number of equations is called an underdetermined system. Consider the

[7]It is known that $V(x)$ cannot be constructed only with an S matrix.

[8]For example, it is used for improving MRI (magnetic resonance imaging) with higher resolution.

[9]In the case of the observation that captured the "shape" of a black hole (called "black hole shadow"), the Fourier components were the observed value and the image data was the output.

following equation:

$$Ax = y\,. \tag{7.11}$$

Here, A is an $n \times m$ matrix, x is an $m$dimensional vector, and y is an $n$dimensional vector. The problem is to find the unknown vector x under the given A and y. As shown in this chapter, the equation system (7.11) is rewritten as a minimization problem[10]

$$\min_x\{|Ax - y|_2^2\} \text{ (for given } y). \tag{7.12}$$

Here the L2-norm $|v|_2$ is defined as $|v|_2 = \sqrt{x_1^2 + x_2^2 + x_3^2 + \cdots}$ for the vector v. Unfortunately, in general, if the number of unknowns is greater than the number of independent equations (which is roughly the amount of the given information), that is, if $m > n$, the system cannot be solved. Such a system is called an underdetermined system. There is not enough information to solve it. The technique described in this chapter is to find

$$\min_x\{|Ax - y|_2^2 - \lambda|x|_2\} \text{ (for given } y \text{ and } \lambda) \tag{7.13}$$

Then you can obtain a plausible solution x. This is called a ridge regression. In this manner the Moore–Penrose inverse was given.

Let us proceed along this direction. For example, let us say that A is sparse and the dimensionality is low, effectively. In this case, the solution x will have many vanishing elements. Therefore, consider the following minimization problem:

$$\min_x\{|Ax - y|_2^2 - \lambda|x|_0\} \text{ (for given } y \text{ and } \lambda). \tag{7.14}$$

Here $|v|_0$ is the number of vanishing elements of the vector v. Thanks to this regularization term, it can still be solved, and a plausible answer can be obtained. And also, this method is effective in the sense that unnecessary zeros are discarded. On the other hand, in order to solve this problem, it is necessary to solve all x elements while checking whether they are 0, which causes a combinatorial explosion and is not practical.

Then, what about using the L1-norm that is halfway between the L2-norm and the L0-norm?

$$\min_x\{|Ax - y|_2^2 - \lambda|x|_1\} \text{ (for given } y \text{ and } \lambda). \tag{7.15}$$

[10]This rewriting is based on the conjugate gradient method [104].

Here, $|v|_1$ is the sum of the absolute values of the components, $|v|_1 = |v_1| + |v_2| + \cdots$ for the vector v. This is a regression called LASSO (least absolute shrinkage and selection operator). This regression is known to have no computational difficulties and to have better properties than the ridge regression.

The problem of solving simultaneous equations can occur in various situations, and one example is when two quantities are related by an integral transformation. This is indeed the case for image sensing and black hole shadows. In the context of physics, the energy spectrum $\rho(\omega)$ and the Green's function $G(\tau)$ are typically related by the following equation: the Fourier transform,

$$G(\tau) = \int d\omega K(\tau, \omega)\rho(\omega). \tag{7.16}$$

Discretizing time τ and energy ω and considering the integral as Riemann sum, this can be regarded as the following equation of a vector $\mathbf{G}$, $\boldsymbol{\rho}$ and a matrix K:

$$\mathbf{G} = K\boldsymbol{\rho}\,. \tag{7.17}$$

We want to know $\rho(\omega)$ from $G(\tau)$. In actual calculations, as for the Green function $G(\tau)$ only about 10 points are known, while as its energy spectrum ρ, we want to obtain more than 1000 points, for example. So, this is an underdetermined system,[11] and you can use LASSO, which we described above.[12] For more details, we suggest that readers look at [105].

[11] In the case of black hole shadows in the EHT (event horizon telescope) project, we know peaks at several values of energy as data, while we want an image that is the inverse Fourier transform. Thus essentially this is the same as the above.

[12] For Green's function, it is necessary to perform a singular value decomposition and move to a base where essential data is easy to see.

Chapter 8
Detection of Phase Transition by Machines

Abstract As an approach to the important question of whether machines can learn the discovery of physics, this chapter examines the question "Can phase transitions be found by deep learning?". Understanding phases is one of the most important subjects in physics. Can machine learning really discover the thermal phase transition in the basic physical system: Ising model?

In this chapter, we explain how to detect a phase transition in the **Ising model** by machine learning. After a brief review of the phase transitions, we will explain phase transition detection using neural networks.

8.1 What is Phase Transition?

Let us review the phase transition of the Ising model, which will be used in this chapter as a target physical system. First, consider a spin variable s_i in the total volume V, and consider the expectation value of the spatial average of the spin variables $M[s] = \frac{1}{V}\sum_i s_i$ in the canonical ensemble, that is, the (spontaneous) **magnetization**. That is given by

$$\langle M \rangle = \frac{1}{Z}\sum_{\{s\}} e^{-\beta H[s]} M[s]\,. \tag{8.1}$$

On the right–hand side, the spin variable s is regarded as a random variable, and the magnetization is defined as the spatial average $M[s]$ of the spin variable under the probability distribution $\frac{1}{Z}e^{-\beta H[s]}$.

If the **Hamiltonian** has a spin inversion symmetry $H[s] = H[-s]$, the magnetization $\langle M \rangle$ always vanishes. This can be shown in the following way. We

A. Tanaka et al., *Deep Learning and Physics*, Mathematical Physics Studies,
https://doi.org/10.1007/978-981-33-6108-9_8

write down the definition of magnetization and massage it as follows:

$$\langle M \rangle = \frac{1}{Z} \sum_{\{s\}} e^{-\beta H[s]} M[s] \tag{8.2}$$

$$= \frac{1}{Z} \sum_{\{-s\}} e^{-\beta H[-s]} M[-s] \tag{8.3}$$

$$= \frac{1}{Z} \sum_{\{s\}} e^{-\beta H[s]} M[-s] \tag{8.4}$$

$$= -\frac{1}{Z} \sum_{\{s\}} e^{-\beta H[s]} M[s] \tag{8.5}$$

$$= - \langle M \rangle . \tag{8.6}$$

So we get the equation $\langle M \rangle = - \langle M \rangle$, but the only solution is $\langle M \rangle = 0$. In the course of obtaining the equation, we used that the state sum variable s is a dummy variable, and that H is symmetric about the flip of s. Also, by definition, $M[-s] = -M[s]$. In the case of Hamiltonians where the spins are aligned and they prefer to be parallel to each other, we can expect spontaneous magnetization to appear, but what made this calculation wrong? In fact, the state sum $\sum_{\{s\}}$ is an infinite sum, so you have to be careful about it. In other words, since the order of the sums cannot be changed arbitrarily, the calculation given above is not always correct: even if H is symmetric with respect to the flip of the sign of s, it could lead to $\langle M \rangle \neq 0$.[1]

More specifically, in the Ising model in two or more dimensions, it is known that $\langle M \rangle \neq 0$ at low temperature while $\langle M \rangle = 0$ at **temperature** higher than a certain value. The temperature region where $\langle M \rangle = 0$ is called the paramagnetic or disordered phase, and the region where $\langle M \rangle \neq 0$ is called the **ferromagnetic phase** or the **ordered phase**. The temperature at which the phases switch is called the phase transition temperature, and the phase change is called the **phase transition**.

---

[1] In order for this to happen, the limit $V \to \infty$ (the **thermodynamic limit**, or the large number limit of the degrees of freedom) is mathematically necessary. In physics, a sufficiently large V can actually be realized. For example, in the Ising model, consider the ground state with all spins pointing up and that with all spins pointing down. A transition between these two states is possible in a finite volume because it has a finite transition probability. Therefore, if the large number limit is not taken, the spin average (the spontaneous magnetization) must be always zero. However, if the transition probability between the states is sufficiently smaller than $\exp(-1/\text{lifetime of the universe})$, it means that there is no practical problem in assuming no transition. Although the volume of a real material is finite, the number of atoms is about the Avogadro number, and in practice, the infinite limit is a good approximation (which is an idealization). To describe the phase transition theoretically using statistical mechanics, one needs to calculate $\lim_{B \to 0} \lim_{V \to \infty} \langle M \rangle$ while keeping the order of the limits, and to find the limit of $\langle M \rangle$ as a function of temperature, where B is the external field that breaks the symmetry (in this case, the magnetic field along the z axis).

The **magnetization** M that appears here is called an **order parameter**, which characterizes the phase transition.

The phase transition phenomenon is not only widely observed in materials such as the magnetization transition and the gas–liquid phase transition of water, but also in the field of elementary particle physics. For example, a vacuum phase transition that changes the nature of the vacuum itself is related to the existence of mass, and it is an important research subject.

Phase transitions are described by finding the symmetry that characterizes the phase and defining the order parameter associated with the symmetry. And typically, the order parameter is derived from a derivative of the free energy. Except for a few models such as the two-dimensional Ising model, the phase transition temperature has not been determined analytically, but can only be determined using numerical calculations. It is also known that the phase transition of topological materials does not have a local order parameter, making it difficult to characterize the phase transition. In such a situation, it is a natural idea to find a phase transition using a neural network. The following describes a study to directly find a phase transition using a neural network.

8.2 Detecting Phase Transition by a Neural Network

When using a neural network for supervised learning, the neural network can do the following:

- When a lot of data $(x, d) =$ (input, answer) are given, it can guess what is the result d' for a new input value x'.

To predict unknown phenomena, a lot of training data must be prepared. Fortunately, in the research in computational physics, a huge amount of data is available through numerical calculations. Particularly in the field of computational physics using the Monte Carlo method, data can be generated almost infinitely (as long as computational resources allow), so the situation is suitable for machine learning.

For example, consider a model with some phase transition in statistical mechanics, and generate its spin configurations by the Monte Carlo method. Can machine learning determine which phase each spin configuration belongs to? The two-dimensional Ising model is the simplest of the models that have a phase transition at finite temperature, and the question of whether a neural network can distinguish the phase transition of the Ising model is positively answered in the literature [106].[2]

[2] Principal component analysis was used to detect a phase transition [107].

First, the training data can be generated relatively easily by using the Metropolis method. Consider a spin on a two-dimensional square lattice. The Hamiltonian H is

$$H[s] = -\sum_{i,j} s_{i,j}(s_{i+1,j} + s_{i,j+1}) . \tag{8.7}$$

Here, the exchange interaction constant is taken as -1, which represents the ferromagnetism. This model is known to undergo a paramagnetic / ferromagnetic phase transition around $T = 2.27$, as mentioned in the column of Chap. 5 (see [74, 75]).

Let us move on to the phase transition detection using neural networks. The spin configuration that serves as the training data is generated by the following procedure:

1. Select a temperature value T.
2. Sample spin configuration s at the temperature T generated by the Metropolis method.
3. If $T < 2.27$, set $d = 0$. If $T > 2.27$, set $d = 1$.
4. Add (s, d) to the training data.

If this step is repeated in the range $T \in [T_{\text{low}}, T_{\text{high}}]$ which includes the phase transition temperature $T = 2.27$, it is possible to generate a training data for a neural network to determine whether a given spin configuration s is in the ordered phase $T < 2.27$ or the disordered phase ($T > 2.27$). In other words, a neural network is designed and trained as a "phase discriminator."

Interesting applications are possible using the neural network trained with this data. Instead of the square lattice Ising model defined by the previous H, it is known that the triangular lattice Ising model

$$H_3[s] = -\sum_{i,j} s_{i,j}(s_{i+1,j} + s_{i,j+1} + s_{i+1,j+1}) \tag{8.8}$$

has a similar phase structure. The phase transition temperature is not $T = 2.27$, but $T = 3.64$. In fact, a neural network trained using H can detect the phase transition of H_3 (Fig. 8.1). We can see that the output of the neural network jumps from 0 to 1 around $T = 3.64$. If this fact holds for other models, we have a new way to detect phase transitions.

The idea of this "phase discriminator" is certainly interesting, but it has its weaknesses when actually discovering unknown phase transitions. The point is that when generating the training data, at least the phase transition point of a similar model must be known. In other words, all of the phase transition detection cannot be covered by the neural network, and it is necessary to know the phase transition point theoretically for a model. This means that the usage is limited. The following simple question comes to mind: can we automatically detect even the phase transition point itself in a single model — in the present case the transition temperature $T = 2.27$?

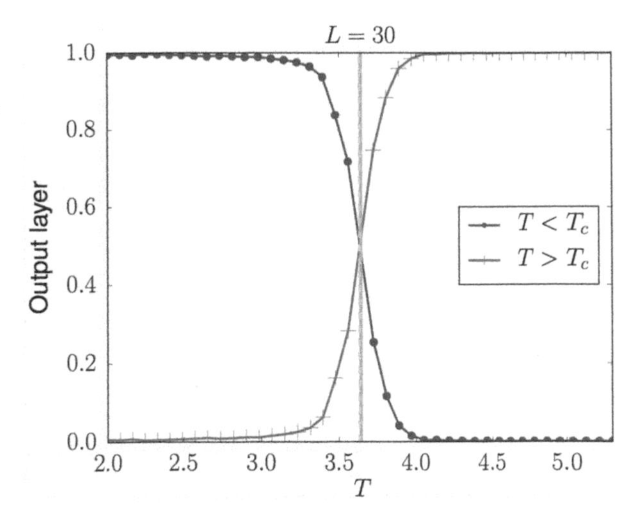

Fig. 8.1 The output of the neural network for each temperature when the configuration of the triangular lattice Ising model H_3 is put as an input data, after the training with the square lattice Ising model H. (This figure is taken from the paper [106])

So let us change our point of view and consider creating a "thermometer" instead of the "phase detector" [108]. In the above-mentioned binary classification, the whole was divided into two classes depending on whether T was 2.27 or less. But this time, without using such prior knowledge of the phase transition temperature, we simply split the whole to N classes. More specifically, the training data is made in the range $[T_{\mathrm{low}}, T_{\mathrm{high}}]$ that includes the phase transition temperature $T = 2.27$, and is classified into N sections: Class 1, Class 2, ..., Class N, depending on the temperature. The error function is chosen to be the softmax entropy.

We visualize the weight J that connects the final layer and the penultimate layer A of the trained neural network. We plot the weights as a pixel image representing the matrix elements of the trained weights J. The horizontal axis is the node of the last layer (the discretized class of the temperature), and the vertical axis is the label m of the components of the layer A. Then, we find that there exists a sudden change in the matrix components near the phase transition temperature (Fig. 8.2). The paper [108] proposed a method for estimating specific phase boundary values from this heat map. As a result, one finds the value of the temperature which is close to $T = 2.27$ (the inverse temperature is $\beta = 0.44\cdots$). Details are given in the original paper, while the results are shown in Table 8.1, and the transition temperature has certainly been detected. This method is interesting because it suggests the possibility of discovering the phase transition phenomenon without knowing the physical properties of the system in advance.

8.3 What the Neural Network Sees

It is also possible to make theoretical considerations as to why this technique can detect phase transition phenomena. Here, we will explain it, based on [109]. First, let us set the number of units in the middle layer to 3 in the following network

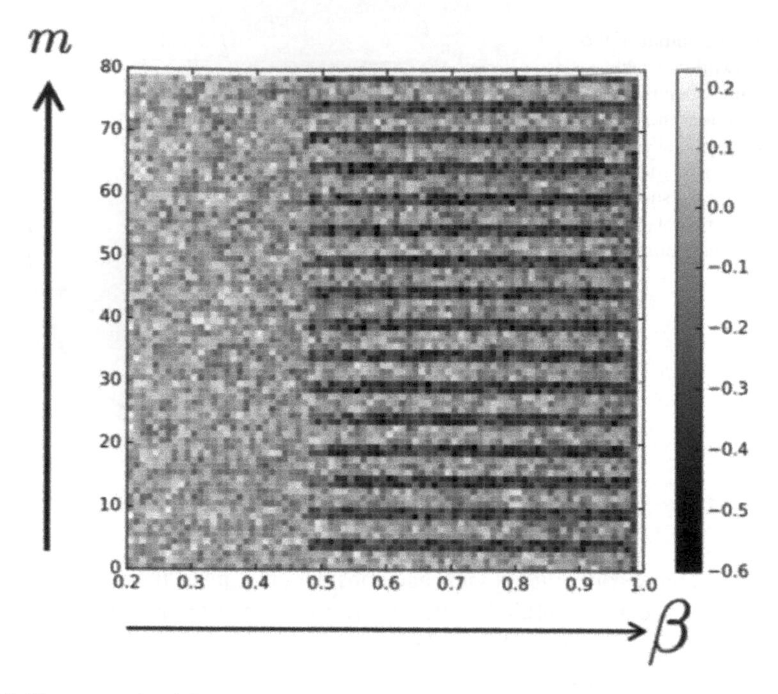

Fig. 8.2 Heat map of weight J adjacent to the final layer of the neural network after the training. (Excerpt from [108])

Table 8.1 The top three rows show the phase transition temperature values detected by the neural network. The values of their inverse temperature β are given. The limit $L \to \infty$ corresponds to an infinite volume, where the exact inverse temperature of the phase transition is known: $\beta_c^{\text{Exact}} = \frac{1}{2}\log(\sqrt{2}+1)$. "CNN" is a convolutional neural network, and "FC" is a fully connected neural network

System size	β_c CNN	β_c FC
8×8	0.478915	0.462494
16×16	0.448562	0.433915
32×32	0.451887	0.415596
$L \to \infty$	$\beta_c^{\text{Exact}} \sim 0.440686$	

of (8.9) that repeats linear and nonlinear transformations twice:

$$f_{\theta,\varphi}(\mathbf{x}) = \boldsymbol{\sigma}(\theta \mathbf{z}), \quad \mathbf{z} = \boldsymbol{\sigma}(\varphi \mathbf{s}). \tag{8.9}$$

Here $\mathbf{s}$ is a vector-like arrangement of the spin configuration s. The activation function $\boldsymbol{\sigma}$ used is a softmax function, and its definition is

$$[\boldsymbol{\sigma}(\mathbf{z})]_I = \frac{e^{z_I}}{\sum_J e^{z_J}}. \tag{8.10}$$

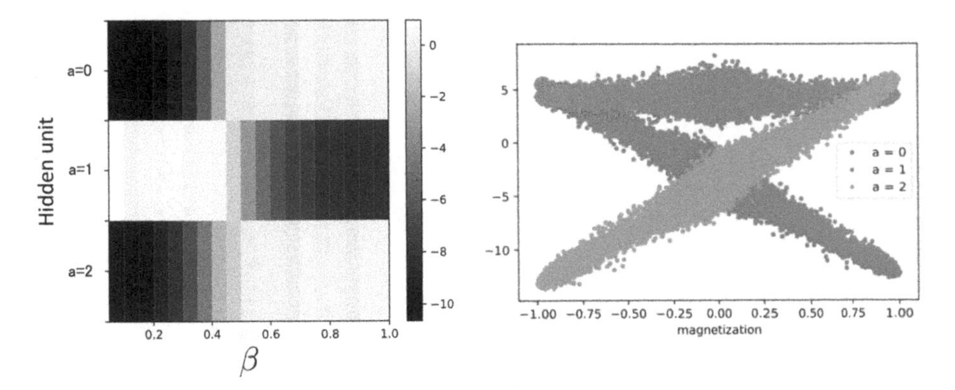

Fig. 8.3 Left: θ^* for the case that the intermediate layer has 3 units. Right: Correlation between each component of $\mathbf{u}$ and the magnetization. (Excerpt from [109])

The output is normalized and so can be regarded as a probability. θ and φ are the parameters to be trained (weights). The results of the numerical experiments are shown on the left of Fig. 8.3. Looking at the θ^* obtained in this case, we can see that the phase transition is again detected.

By the way, what kind of information is hidden in the following intermediate three-dimensional vector?

$$\mathbf{u} = \begin{pmatrix} u_{\text{red}} \\ u_{\text{green}} \\ u_{\text{blue}} \end{pmatrix} = \varphi^* \mathbf{x} . \tag{8.11}$$

The vector components are plotted for various spin configurations (the magnetization $\langle M \rangle$ on the horizontal axis and $\mathbf{u}$ on the vertical axis). See the right panel of Fig. 8.3. As you can see, there exists a clear correlation, so we can conclude that the order parameter of the phase transition has been learned at the middle layer.

Once you deepen the network, you can actually confirm that even the information about the internal energy is directly embedded. This suggests a theoretical connection between neural networks and statistical systems. Interested readers should refer to the paper [109], and a related reference [110].

Chapter 9
Dynamical Systems and Neural Networks

Abstract Neural networks are a way of expressing a variety of nonlinear functions, but can also be thought of as waves of information propagating between layers. In this chapter, we show that such multi-layer propagation can be interpreted as the time evolution of dynamical systems, and hence of Hamiltonian systems, and look at the close relationship between the fundamental concept of "time evolution" in physics and deep neural networks.

If a variety of physical systems can be represented using neural networks, it will greatly open the possibility of applying the systems to machine learning for analysis. In physics, differential equations are the basic equations due to the concept of locality and causality, so it is necessary to find out what differential equations allow the representation of neural networks. In machine learning, there are so many types of neural networks, and a new network structure is proposed mainly from the viewpoint of improving learning efficiency, so the correspondence to **differential equations** is not clear. In other words, the network obtained by discretizing the differential equation has a different intention from the neural network for machine learning and deep learning.

In this chapter, we review the relationship between differential equations and neural networks, and discuss, in particular, Hamiltonian dynamical systems, and which Hamiltonians allow the structure of a typical neural network.

9.1 Differential Equations and Neural Networks

First, let us define what we call "a typical neural network" in this chapter. Consider a neural network given in Fig. 9.1. The layers are arranged from left to right, and each layer has a vector x_i of the same dimension (the subscript i labels the elements of the vector). Between layers, the linear transformation $x_i \rightarrow J_{ij}x_j$ and the local nonlinear transformation $x_i \rightarrow \sigma(x_i)$ by the activation function act in order. With

A. Tanaka et al., *Deep Learning and Physics*, Mathematical Physics Studies,
https://doi.org/10.1007/978-981-33-6108-9_9

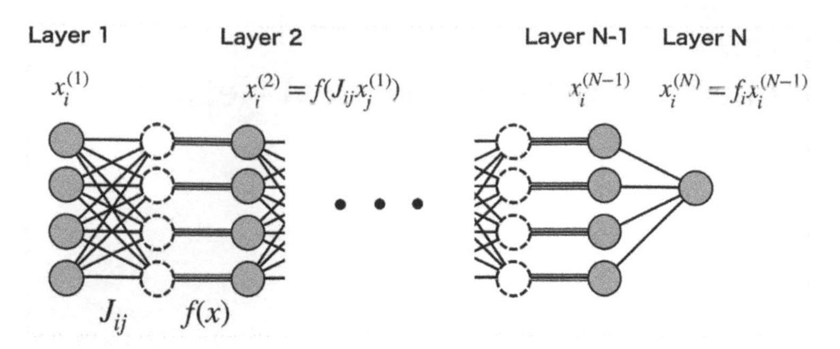

Fig. 9.1 Typical deep neural network. The solid line represents the linear transformation multiplied by the matrix J, and the triple line represents the nonlinear transformation (activation function)

this stacking, the final output of the neural network is written as

$$y(x^{(1)}) = f_i \sigma(J_{ij}^{(N-1)} \sigma(J_{jk}^{(N-2)} \cdots \sigma(J_{lm}^{(1)} x_m^{(1)}))) . \tag{9.1}$$

To reiterate, learning means, by changing the network variables $(f_i, J_{ij}^{(n)})$ $(n = 1, 2, \cdots, N-1)$, to minimize the following **error function**:

$$L \equiv \sum_{\text{data}} \left| y(\overline{x}^{(1)}) - \overline{y} \right| + L_{\text{reg}}(J) . \tag{9.2}$$

Here the sum runs over the entire set of training data pairs $\{(\overline{x}^{(1)}, \overline{y})\}$. The input data $\overline{x}^{(1)}$ is put to the first layer, and $\overline{y}$ is the correct output data that should be output from the last layer. The additional term L_{reg} is called the regularization term and is introduced to control learning.

Now let us look at the relationship between neural networks and differential equations. In 2016, a deep neural network called **ResNet** (Residual Network, **residual neural network**) was proposed [24]. It is known as an efficient network where learning progresses even with a very large number of layers. In this method, a detour is provided in the neural network, and the detour is joined without any modification:

$$x_i^{(n+1)} = f(J_{ij} x_j^{(n)}) + x_i^{(n)} . \tag{9.3}$$

The first term on the right-hand side is the typical neural network, but the second term is a "skip connection," the detour. By adding such terms, it has been found that

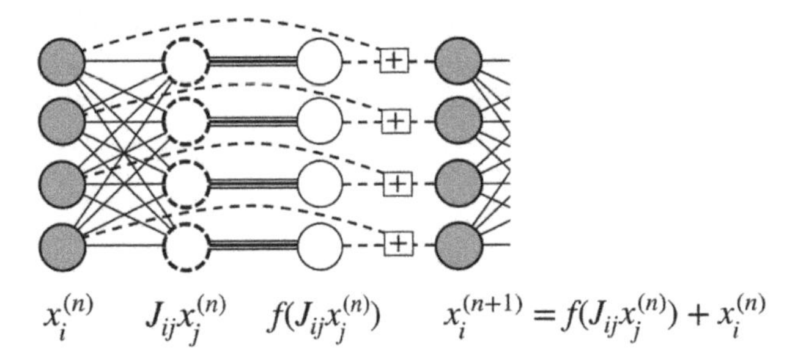

Fig. 9.2 Conceptual diagram of the ResNet. The effect of adding $x_i^{(n)}$, represented by a dotted line, is combined with the structure of the typical deep neural network represented in Fig. 9.1. The box surrounding the "+" represents a linear transformation that adds the inputs

learning progresses even when the network goes deeper.[1] It is thought that the reason may be that error backpropagation proceeds efficiently. The conceptual diagram of ResNet is shown in Fig. 9.2.

This ResNet can be interpreted as a discretized version of a differential equation [112]. Let us consider an equation that determines the time evolution of a dynamical system,

$$\dot{x}_i(t) = f_i(x_j(t)) . \tag{9.5}$$

If we discretize the time coordinate, we find

$$x_i(t_{n+1}) = x_i(t_n) + (\Delta t) f_i(x_j(t_n)) , \quad t_{n+1} = t_n + \Delta t . \tag{9.6}$$

The ResNet (9.3) has this form. That is, the deepened ResNet becomes a continuous time evolution equation if an appropriate limit on the size of the activation function (this size is equivalent to Δt and is regarded as a hyperparameter for learning) is taken.[2]

Note that not all differential equations in dynamical systems can be written like ResNet. If x has several components, that is, if there are more than one unit, you cannot generally write the differential equation like the ResNet. In order to form a neural network, the nonlinear term must be in the form of an activation function of (9.3), that is, $f(J_{ij}x_j)$. Any $f_i(x(t))$ that defines a dynamical system is not

[1] Even before the advent of ResNet, the possibility of a detour was considered. It is called a highway network [111] and has the following form:

$$x^{(n+1)} = T\left(\tilde{J}x^{(n)}\right) f\left(Jx^{(n)}\right) + \left(1 - T\left(\tilde{J}x^{(n)}\right)\right) x^{(n)} . \tag{9.4}$$

Here, $T(\tilde{J}x^{(n)})$, which is the variable parameter of the network, determines the ratio of the amount to be sent to the detour. If $T(\tilde{J}x^{(n)})$ is a constant and $T = 1/2$, the ResNet (9.3) can be obtained.

[2] The continuous limit of deepening is discussed in [113] from the viewpoint of data assimilation.

always in that form. We will look at these subtleties in more detail later in the case of Hamiltonian dynamical systems.

Now, as a further generalization of ResNet, there is a neural network called **RevNet** (reversible residual network) [114]. The RevNet is again a residual learning, but has a symmetric form:

$$x_i^{(n+1)} = f(J_{ij} y_j^{(n)}) + x_i^{(n)} , \tag{9.7}$$

$$y_i^{(n+1)} = g(J_{ij} x_j^{(n+1)}) + y_i^{(n)} . \tag{9.8}$$

With the same analogy, it can be seen that this neural network is a discretized version of the following dynamical system:

$$\dot{x}(t) = f(y(t)) , \quad \dot{y}(t) = g(x(t)) . \tag{9.9}$$

Reversible means that you can return to the input layer in order from the data output value at the last layer. If you look closely at the right-hand side of (9.8), you find $x_j^{(n+1)}$ instead of $x_j^{(n)}$. With this, if output data $(y_i^{(n+1)}, x_i^{(n+1)})$ is given, first $y_i^{(n)}$ is obtained through (9.8), then (9.7) returns $x_i^{(n)}$, and so on. Being able to return is actually related to the amount of memory used for learning. In the normal neural network backpropagation method, the values of the weight at each layer must be stored in memory. However, in the case of a reversible neural network, the values of the weight at each layer can be recalculated from the data of the last layer, so there is no need to store the weight in memory, and efficient learning concerning memory consumption is provided.

Furthermore, being reversible is deeply related to differential equations in dynamical systems. In dynamical systems, a characteristic behavior called **chaos** is one of the important research subjects. Chaos is the sensitivity to the initial values. In terms of neural networks, the output value changes completely even if the initial input data is slightly perturbed. A neural network obtained by a discretization of a chaotic dynamical system is susceptible to perturbation of the input data, that is, learning is considered to be difficult. A neural network based on a dynamical system without chaos is called a "stable neural network" [115]. On the other hand, in a reversible neural network, a similar problem can occur if there is chaotic behavior when considering propagation in the reverse direction. In this case, rather, it corresponds to the situation where the initial value difference completely collapses before reaching the final layer (the output data does not change due to the difference in the input data). It is related to the existence of an **attractor**. The degree of chaos in a dynamical system is measured by a constant called the **Lyapunov exponent**. The system is chaotic when it is positive, while the system is "collapsing" when it is negative. Dynamical systems in which the Lyapunov exponent has no real part are appropriate for reversible neural networks [116].

In addition, there are neural networks created by imitating Hamilton equations[3] and also those that are made from second-order differential equations instead of the first order. In this way, the skip connection that appeared to deepen and efficiently train neural networks has a convenient form to interpret the neural network as a discretized version of differential equations.[4]

9.2 Representation of Hamiltonian Dynamical System

It is difficult to represent the time evolution of any Hamiltonian dynamical system by a neural network. However, for a limited class of Hamiltonians, representation of neural networks using local activation functions can be easily obtained. Let us look at this.[5]

When a **Hamiltonian** $H(p, q)$ is given, the time evolution of the system is determined by the following **Hamilton equation**:

$$\dot{q} = \frac{\partial H}{\partial p}, \quad \dot{p} = -\frac{\partial H}{\partial q}. \tag{9.12}$$

Here, for simplicity, we consider a one-dimensional system, that is, one $p(t)$ and one $q(t)$, but generalization to a multidimensional system is easy.[6]

First, let us try the simplest interpretation. It takes the dimension of the vector of each layer of the neural network as 2, equates it with $(q(t), p(t))$. Then we discretize the t direction, and regard it as the depth direction in which the layers are stacking. This method unfortunately shows that only free Hamiltonians (namely, Hamiltonians consisting only of second-order polynomials in p and q) allow a neural network representation, as we will see below.

---

[3]For example, reference [115] gives a neural network that discretizes the following differential equation based on the Hamiltonian dynamical system:

$$\dot{y}(t) = \sigma(k(t)z(t) + b(t)), \tag{9.10}$$

$$\dot{z}(t) = -\sigma(k(t)y(t) + b(t)). \tag{9.11}$$

Here, $k(t)$ and $b(t)$ are parts corresponding to weight and bias, and σ is an activation function. As you can see from the minus sign in front of the right-hand side of the second equation, this is a neural network-like differential equation inspired by the Hamilton equation. (However, if you look carefully, this differential equation is not derived from any Hamiltonian.)

[4]Also, the method of using the ordinary differential equation itself as a neural network has been studied [117].

[5]The result of this section is based on [118].

[6]Identification of Hamiltonian dynamical systems with neural networks in other ways can be found, for example, in the literature [119]. Chapter 12 introduces the neural network representation of nonlinear ordinary differential equations in string theory applications.

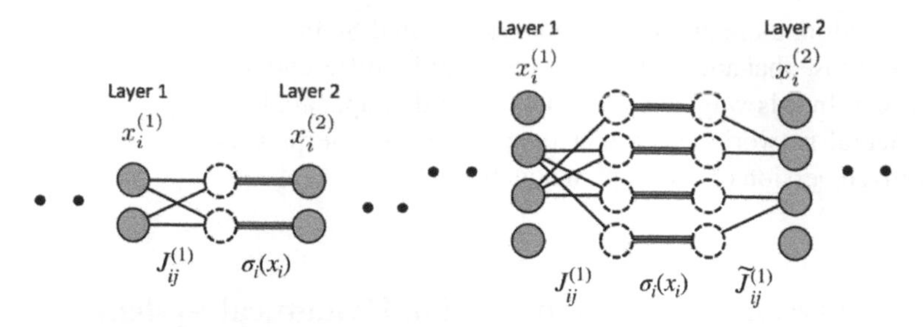

Fig. 9.3 Left: A simple network that considers canonical variables as units of a neural network and views the transformation between layers as discrete time translation. Right: Neural network extended to represent more general Hamiltonian dynamical systems

A neural network that identifies the discrete-time translation $t \to t + \Delta t$ with the transformation between layers can be written as

$$\begin{aligned} q(t + \Delta t) &= \sigma_1(J_{11}q(t) + J_{12}p(t)), \\ p(t + \Delta t) &= \sigma_2(J_{21}q(t) + J_{22}p(t)). \end{aligned} \tag{9.13}$$

That is, the linear transformation J and the nonlinear local transformation σ are performed successively. Here, "local" means that the argument of σ_1 is only the value of the first unit, and the argument of σ_2 is only the value of the second unit. This network is shown in Fig. 9.3 left. Here, the units $x_1^{(n)}$ and $x_2^{(n)}$ are directly identified with $q(t)$ and $p(t)$. The time t is discretized, and the interval is Δt; this is expressed as $t = n\Delta t$.

Then, can the Hamilton equation (9.12) be written as a neural network (9.13)? First, in order for (9.13) to be interpreted as a discretized version of the differential equation in continuous time, (9.13) must have a consistent limit $\Delta t \to 0$. The following conditions are required for the weight J and the activation function σ:

$$J_{11} = 1 + O(\Delta t), \quad J_{22} = 1 + O(\Delta t), \tag{9.14}$$

$$J_{12} = O(\Delta t), \quad J_{21} = O(\Delta t), \tag{9.15}$$

$$\sigma(x) = x + O(\Delta t). \tag{9.16}$$

To satisfy these, let us assume the following:

$$J_{ij} = \delta_{ij} + w_{ij}\Delta t, \quad \sigma_i(x) = x + g_i(x)\Delta t. \tag{9.17}$$

Here w_{ij} $(i, j = 1, 2)$ is the weight (constant parameters) and $g_i(x)$ $(i = 1, 2)$ is a nonlinear function. Substituting these into (9.13) and taking the limit of $\Delta t \to 0$

yields the following formula:

$$\dot{q} = w_{11}q + w_{12}p + g_1(q)\,, \tag{9.18}$$

$$\dot{p} = w_{21}q + w_{22}p + g_2(p)\,. \tag{9.19}$$

In order for them to be Hamilton equations (9.12), their right-hand sides must satisfy the symplectic relation:

$$\frac{\partial}{\partial q}(w_{11}q + w_{12}p + g_1(q)) + \frac{\partial}{\partial p}(w_{21}q + w_{22}p + g_2(p)) = 0\,. \tag{9.20}$$

However, this equation does not allow any nonlinear $g_i(x)$. Therefore, simply equating the unit vector with (p, q) yields only a linear Hamilton equation.

So, let us use some ingenuity. We extend the relationship between units and canonical variables, and also extend the relationship between time evolution and transformation between layers. We shall assume the following relationship:

$$x_i(t + \Delta t) = \widetilde{J}_{ij}\sigma_j(J_{jk}x_k(t))\,. \tag{9.21}$$

This differs from (9.13) in two ways: First, $x_1 = q$ and $x_2 = p$ are the same as before, but we add x_0 and x_3, and set $i, j, k = 0, 1, 2, 3$. Second, it introduces $\widetilde{J}$. This second point does not change the neural network. We simply split J between the layers into two linear transformations and rewrote the first linear transformation as a transformation working at the previous layer. Thus, the following combination, the linear translation $J \to$ nonlinear local transformation σ $\to$ the linear transformation $\widetilde{J}$, is regarded as a Δt time translation.

In the neural network expanded like this, we choose sparse weights and local activation functions as follows:

$$J = \begin{pmatrix} 0 & 0 & v & 0 \\ 0 & 1 + w_{11}\Delta t & w_{12}\Delta t & 0 \\ 0 & w_{21}\Delta t & 1 + w_{22}\Delta t & 0 \\ 0 & u & 0 & 0 \end{pmatrix}, \quad \widetilde{J} = \begin{pmatrix} 0 & 0 & 0 & 0 \\ \lambda_1 & 1 & 0 & 0 \\ 0 & 0 & 1 & \lambda_2 \\ 0 & 0 & 0 & 0 \end{pmatrix}, \tag{9.22}$$

$$\begin{pmatrix} \sigma_0(x_0) \\ \sigma_1(x_1) \\ \sigma_2(x_2) \\ \sigma_3(x_3) \end{pmatrix} = \begin{pmatrix} f(x_0)\Delta t \\ 1 \\ 1 \\ g(x_3)\Delta t \end{pmatrix}. \tag{9.23}$$

Here (u, v, w_{ij}) $(i, j = 1, 2)$ is a weight constant. The conceptual picture of this neural network is shown in the right panel of Fig. 9.3.

Using this definition of time translation, the transformation between layers is

$$\dot{q} = w_{11}q + w_{12}p + \lambda_1 f(vp), \tag{9.24}$$

$$\dot{p} = w_{21}q + w_{22}p + \lambda_2 g(uq). \tag{9.25}$$

The condition of the symplectic structure of the Hamilton equation is

$$w_{11} + w_{22} = 0 \tag{9.26}$$

which can be easily satisfied. The corresponding Hamiltonian is

$$H = w_{11}pq + \frac{1}{2}w_{12}p^2 - \frac{1}{2}w_{21}q^2 + \frac{\lambda_1}{v}F(vp) - \frac{\lambda_2}{u}G(uq), \tag{9.27}$$

and we chose here

$$F'(x_0) = f(x_0), \quad G'(x_3) = g(x_3). \tag{9.28}$$

This is a nonlinear Hamiltonian with a deep neural network representation.

As an example, if we choose

$$w_{11} = w_{21} = 0, \quad w_{12} = 1/m, \quad \lambda_1 = 0, \quad \lambda_2 = 1, \quad u = 1, \tag{9.29}$$

the corresponding Hamiltonian is the Hamiltonian of a nonrelativistic particle moving in an arbitrary potential:

$$H = \frac{1}{2m}p^2 - G(q). \tag{9.30}$$

It is easy to see that if you devise $F(p)$, you can also create a relativistic particle Hamiltonian.

In order to construct, using a neural network, a general nonlinear Hamiltonian in which both p and q are included as nonlinear functions, a network configuration using a basis of linear combinations for the whole nonlinearity is required. Although not described in detail here, various types of Hamiltonians can be built by generalizing the network as described above.[7]

In this chapter, we have seen that the time evolution of differential equations and Hamilton equations can be reconstructed as data propagation over neural networks. By using the method introduced here, the time evolution of physical systems can be directly applied to the deep learning scheme. The main issue of machine learning is generalization and solving inverse problems. If the problem of time evolution of a

[7] Nonlinear Schrödinger equations can be constructed in a similar way.

physical system is an inverse problem or is based on application to unknown data, the method in this chapter may be effective.

In Chap. 12, we explain a solution by deep learning, taking an example where an evolution problem of a physical system is an inverse problem. In other words, it is a problem to determine the evolution equation itself, when the evolution equation is unknown and the data before and after the evolution is given. Determining a system of equations is the most essential part of physics, and using machine learning techniques on such an aspect can sometimes help to analyze it effectively.

Chapter 10
Spinglass and Neural Networks

Abstract To begin with, neural networks are based on neural circuits formed by neurons in human brains. One of the important mechanisms of the brain is memory. The Hopfield model, which explains the mechanism of memory in terms of physics, is a bridge between physics and neural networks. In this chapter, we explain the Hopfield model and investigate the relationship between machine learning and spin glass, which is still a rich subject in condensed matter physics.

The similarity between the neural network and the physical system cannot be described without the **Hopfield model** [120]. This model explains the mechanism of brain memory by a system in which many **spins** are coupled.[1] Let us consider that many particles with spins gather and there are various interactions among the spins. There appear many metastable states at low energy, and they degenerate. Such a system is called **spinglass**. It is called spinglass because glass can take many states instead of atoms being lined up in order like a solid. Such a classical spinglass system can be considered as a kind of a neural network.

Linking neural networks to our familiar physical systems is one of the subjects of this book. It is interesting to see that the spinglass system, which plays a leading role in statistical physics, is related to neural networks, and it is one major intersection that links physics and machine learning. Hopfield has modeled human **memory** (information storage) using some collective dynamics of spinglasses, as we will see below. This gives the idea that long-term memory can be considered as an attractor of dynamical systems. Historically, from this perspective, the chaos of brains, that is, the understanding of the brain as a dynamical system, has evolved. In this chapter, we will return to the idea of Hopfield and look at the relationship between spinglasses and neural networks.

[1] S. Amari published a similar model ten years earlier [121], and the Hopfield's model is also known as the Amari–Hopfield model.

A. Tanaka et al., *Deep Learning and Physics*, Mathematical Physics Studies,
https://doi.org/10.1007/978-981-33-6108-9_10

10.1 Hopfield Model and Spinglass

Let us introduce the Hopfield model, a brain model of long-term memory and association. The neural circuit is composed of the mutual connection of **neural cells** (**neurons**). Neurons are connected at synapses, and neurons can have firing and non-firing states. When one neuron is firing, signals are transmitted through the synapse to the next neuron. The next neuron fires when the weighted sum of all incoming electrical signals exceeds a certain threshold.

We model such a system as follows. In a system in which N neurons are collected, the state of each neuron is $\{s_i\}$ $(i = 1, 2, \cdots, N)$, and the firing state of the i–th neuron is defined as $s_i = 1$ and the non-firing state as $s_i = 0$. Then, write the synaptic strength between the i–th neuron and the j–th neuron as J_{ij}. The rule that determines that the input changes the state s_i to the state s_i' is

$$s_i \to s_i' \equiv \theta\left(\sum_j J_{ij} s_j - h_i\right). \tag{10.1}$$

Here, h_i is the threshold, and $\theta(x)$ is the step function:

$$\theta(x) = \begin{cases} 1 \ (x > 0), \\ 0 \ (x \leq 0). \end{cases} \tag{10.2}$$

Therefore, when the weighted sum of input signals $\sum_j J_{ij} s_j$ exceeds the threshold h_i, it fires: $s_i' = 1$.

The rule of the Hopfield model (10.1) appears to be the same as those of the neural networks with the inter-layer data propagation, as you look back at the neural networks that have appeared so far, if you regard the step function as an activation function. So how is the Hopfield model different from deep learning?

First, the deep neural network used for ordinary deep learning has a hierarchical layer structure, and adopts a rule of successively propagating from the input layer to the output layer. On the other hand, in the Hopfield model, one neuron is extracted at random and only that neuron is updated with the rule (10.1). Then one repeats the operation of taking out randomly again. This is based on the idea that updates in the real brain are not perfectly synchronized in time. Therefore, the Hopfield model does not have the framework of propagation through the layers.[2]

Related to this, the Hopfield model assumes the following:

$$J_{ij} = J_{ji}. \tag{10.3}$$

[2]The Hopfield model is similar to the Boltzmann machine, and a Boltzmann machine that does not allow intra-layer coupling is called a restricted Boltzmann machine (RBM). See Chap. 6.

Unlike the connection of neurons in the brain, the connection between neurons is assumed to be bidirectional. This emphasizes the similarity with spin systems.

Now, in order to see the relationship with the spinglass, redefine the degrees of freedom as follows:

$$S_i \equiv 2s_i - 1 . \tag{10.4}$$

Then, $S_i = \pm 1$ can be interpreted as up and down spins.[3] Furthermore, suppose a virtual $(N+1)$–th spin which only takes $S_{N+1} = +1$. Then, the update rule (10.1) is put together in the following concise form:

$$S_i \to S_i' \equiv \text{sgn}\left(\sum_{i=1}^{N+1} J_{ij} S_j\right) . \tag{10.5}$$

Here we have defined[4]

$$J_{i,N+1} = J_{N+1,i} \equiv \sum_{j=1}^{N} J_{ij} - 2h_i . \tag{10.6}$$

And, $\text{sgn}(x)$ is a sign function defined as

$$\text{sgn}(x) = \begin{cases} 1 & (x > 0) , \\ -1 & (x \le 0) . \end{cases} \tag{10.7}$$

It is useful to remember that at $x \neq 0$ it can be written as $\text{sgn}(x) = x/|x|$.

Now let us consider how the whole spin behaves when the rule (10.1) is applied. In a spin system, we define the following fundamental function, which is the potential energy (interaction energy between spins):

$$E \equiv -\frac{1}{2} \sum_{i,j} J_{ij} S_i S_j . \tag{10.8}$$

One can show that this function is a **Lyapunov function** as follows. The Lyapunov function is a function which changes monotonically under the evolution. It is used to find the stability of the equilibrium point of a dynamical system, and can be thought

[3]In quantum spin systems, this binary system can be considered as spin 1/2, but here we consider classical systems only.

[4]If the self-coupling $J_{N+1,N+1}$ is taken sufficiently large positive, $S_{N+1} = +1$ will be retained in the updates.

of as physically equivalent to the energy of the system. The change in E under the rule is

$$\Delta E = -\sum_{i,j} J_{ij} S_j \Delta S_i \,. \tag{10.9}$$

If $\Delta S_i = (+1) - (-1) = 2$, then $\sum_{i,j} J_{ij} S_j > 0$, so the contribution is $\Delta E < 0$. If $\Delta S_i = (-1) - (+1) = -2$, then $\sum_{i,j} J_{ij} S_j < 0$, so again the contribution is negative, namely $\Delta E < 0$. If $\Delta S_i = 0$, the contribution has no effect on ΔE. Therefore, we can show

$$\Delta E \leq 0. \tag{10.10}$$

That is, E is invariant or monotonically decreasing and satisfies the properties of any Lyapunov function.

The meaning that the function E plays the role of energy should be understood in the broad sense that the system proceeds in the direction of decreasing energy.[5] In this sense, the Hopfield model is similar to the potential energy part of many-body spin systems. In fact, the Hopfield model was the historical origin of the Boltzmann machine described in Chap. 6.

What is the spin configuration that minimizes or extremizes the potential energy (10.9)? Considering a general J_{ij}, we can see that a very large number of configurations achieve a local minimum. Such a system is called a spinglass. For example, if spins are arranged at equal intervals on a one-dimensional line, and only adjacent spins have a bond of $J_{i,i+1} = J(>0)$, the lowest energy state realized is ferromagnetic, that is, $S_i = +1$ for all i's. In this case, there is no degeneration. On the other hand, for example, considering only three spins, if $J_{12} = J_{13} = -J_{23} = J(>0)$, the lowest energy state has six patterns, thus there is a sixfold degeneracy. The three spins are in a state of three-way deadlock, and if you try to lower any one of the interaction energies, some other will increase. The system in which such a state is realized is called the system with "frustration." If you allow a general coupling as J whose value can be a positive or negative general value, you will have a frustrated system.

In summary, in the Hopfield model, when the firing state of each neuron evolves according to the rule (10.1), the state finally reaches various states which are quite degenerate.

[5]Energy is conserved if it has a time-translation invariant Hamiltonian. In that sense, the Lyapunov function is not equivalent to energy. However, for example, if the Hamiltonian system is in contact with an external heatbath and the temperature of the heatbath is low, the energy drops monotonically in time.

10.2 Memory and Attractor

In the Hopfield model, "memory" refers to the set of spin values $\{S_i = a_i\}$. The information of the memory is stored in a **synapse** J_{ij}. Assume that the synapses have the following values ($\alpha > 0$), respectively, by a certain mechanism which we explain later:

$$J_{ij} = \alpha a_i a_j \,. \tag{10.11}$$

Then the memory $\{S_i = a_i\}$ is a fixed point under the update rule (10.1):

$$S_i' = \mathrm{sgn}\left(\sum_j J_{ij} a_j\right) = \mathrm{sgn}\left(\sum_j \alpha a_i a_j a_j\right) = \mathrm{sgn}\left(\alpha(N+1)a_i\right) = a_i \,. \tag{10.12}$$

Therefore, once the spin state falls into $\{S_i = a_i\}$, it will stay there.

Is this state stable? When the state changes a little from the fixed point by some external perturbation, will it return to the original fixed point? Let us evaluate the Lyapunov function (10.9),

$$E = -\frac{\alpha}{2}\sum_{ij} a_i a_i a_j a_j = -\frac{\alpha}{2}(N+1)^2 \,. \tag{10.13}$$

This has a very large negative value. In fact, the value of $O(N^2)$ is the maximum magnitude that the Lyapunov function can have, and the memory $\{S_i = a_i\}$ is at the bottom of a very deep valley, so is a stable fixed point. In other words, when starting from an arbitrary firing state, the state is updated according to the rule (10.1), and finally reaches a stable fixed point $\{S_i = a_i\}$ and stops moving. Such a fixed point is called an **attractor**, from the viewpoint of dynamical systems. Reaching the attractor is interpreted as a mechanism of "recalling" memory.

For an important memory the Lyapunov function must fall into deeper valleys. And in addition, there are many shallow valleys because of the frustration. Even if it gets stuck in such a shallow valley, it is expected that various external perturbations will finally lead to the deep valley.

Of course, there are many patterns to remember, not just a single pattern $\{S_i = a_i\}$. So, let us consider a situation where you want to store M kinds of patterns $\{S_i = a_i^{(m)}\}$ ($m = 1, 2, \cdots, M$). Then the appropriate coupling is

$$J_{ij} = \frac{\alpha}{M}\sum_m a_i^{(m)} a_j^{(m)} \,. \tag{10.14}$$

In particular, it is assumed that the patterns of memory are orthogonal to each other:

$$\sum_i a_i^{(m)} a_i^{(n)} = (N+1)\delta_{m,n} \,. \tag{10.15}$$

Then, as before, you can see that each memory $\{S_i = a_i^{(m)}\}$ is a fixed point under the update rule (10.1),

$$\begin{aligned} S_i' &= \mathrm{sgn}\left(\sum_j J_{ij} a_j^{(m)}\right) \\ &= \mathrm{sgn}\left(\sum_n \sum_j \frac{\alpha}{M} a_i^{(n)} a_j^{(n)} a_j^{(m)}\right) \\ &= \mathrm{sgn}\left(\frac{\alpha(N+1)}{M} a_i^{(m)}\right) \\ &= a_i^{(m)} . \end{aligned} \tag{10.16}$$

Here, the orthogonal condition of memory (10.15) was used. Also, we can evaluate the Lyapunov function (10.9) about the memory $\{S_i = a_i^{(m)}\}$ as

$$E = -\frac{\alpha}{2M} \sum_{i,j} \left(\sum_n a_i^{(n)} a_j^{(n)}\right) a_i^{(m)} a_j^{(m)} = -\frac{\alpha}{2M}(N+1)^2 . \tag{10.17}$$

After all, each memory is at the bottom of a deep valley of $O(N^2)$.

If you start updating from any initial conditions, you are expected to eventually reach a "near" deep valley. This can be interpreted as a mechanism whereby past memories are awakened when a near phenomenon is perceived. Here, âĂIJnearâĂİ specifically depends on the details of the structure of the Lyapunov function. Due to the orthogonality condition of memory (10.15), it is expected that the "near" may mean that the inner product with a pattern $\{S_i = a_i^{(m)}\}$ is large. This is because the evaluation method of the Lyapunov function is the inner product, as in (10.17).

So far, we have seen that when the coupling J_{ij} takes the value (10.11), a deep valley is realized, which behaves as a stable fixed point. Then, how is the value (10.11) itself realized? In supervised machine learning, neural networks are trained by updating weights using the gradient descent method, while the Hopfield model does not look at the correlation between an input and an output. The realization of (10.11) is thought to be performed by a theory generally called Hebbian learning theory. **Hebb's rule** is a rule that synapses are strengthened when both of the neurons they connect are firing, while they will attenuate when the neurons did not fire. From this idea, for example, an equation like the following is expected:

$$\mu \frac{d}{dt} J_{ij}(t) = -J_{ij}(t) + \alpha a_i a_j . \tag{10.18}$$

The first term on the right-hand side is the attenuation term. The second term on the right-hand side is a reinforcement term that depends on the state of both ends connected by the synapse. When the memory pattern $\{S_i = a_i^{(m)}\}$ is input from outside for a long period of time, J will be updated according to this kind of equation. Since the equation is written to converge exponentially to (10.11), the shape of the reinforcement term provides the appropriate weight.

10.3 Synchronization and Layering

Updating only randomly selected spins according to the rule (10.1) is not similar to that of a multilayer deep neural network. However, modifying the Hopfield model such that it updates all spins at the same time makes it possible to incorporate a hierarchical structure.

Since we update the state of all spins at once, we shall label the number of steps in that synchronized update as n:

$$S_i(n) \to S_i(n+1) \equiv \mathrm{sgn}\left(\sum_{i=1}^{N+1} J_{ij} S_j(n)\right) . \tag{10.19}$$

Then, as a Lyapunov function, it is known that the following choice works well:

$$E(n) \equiv -\frac{1}{2}\sum_{ij} J_{ij} S_i(n+1) S_j(n) . \tag{10.20}$$

In fact, with a quantity that shows how $E(n)$ changes with updates,

$$\Delta E(n) \equiv E(n+1) - E(n) , \tag{10.21}$$

we can show that it is always negative or zero, as follows. First, let us massage $\Delta E(n)$:

$$\begin{aligned}
\Delta E(n) &= -\frac{1}{2}\sum_{ij} J_{ij} \left(S_i(n+2) - S_i(n)\right) S_j(n+1) \\
&= -\frac{1}{2}\sum_{i} \left(S_i(n+2) - S_i(n)\right) \sum_{j} J_{ij} S_j(n+1) \\
&= -\frac{1}{2}\sum_{i} \left(S_i(n+2) - S_i(n)\right) \frac{S_i(n+2)}{|\sum_j J_{ij} S_j(n+1)|} .
\end{aligned} \tag{10.22}$$

For the last line, we used $\text{sgn}(x) = x/|x|$. In this last line, using

$$(S_i(n+2) - S_i(n))\, S_i(n+2) = \quad 0 \quad \text{or} \quad 2\,, \tag{10.23}$$

the following has been proven as expected:

$$\Delta E(n) \leq 0. \tag{10.24}$$

In other words, $E(n)$ is a Lyapunov function, and it is shown that $E(n)$ evolves to a smaller value by the updates of the system.

When J_{ij} satisfies (10.11), it can be shown by the same logic as before that the fixed point is $\{S_i = a_i\}$. At this fixed point, $E(n)$ takes the value of $O(N^2)$, so it is an attractor. The difference from the case without synchronization is that two types of attractor behavior are allowed due to the difference of the Lyapunov function. The first type is the same as the previous fixed point, but the second type can be

$$S_i(n+2) = S_i(n)\,. \tag{10.25}$$

This condition yields $\Delta E(n) = 0$, which results in a stable orbit. In dynamical systems, such a cycle is called a **limit cycle**. Therefore, we conclude that, in addition to the usual fixed point, a limit cycle of period 2 is allowed as a stable trajectory.

Now, the synchronous Hopfield model can be rewritten as a hierarchical network. In other words, by regarding the updated label n as a layer label, the Hopfield model is copied by an arbitrary number of layers. If you prepare multiple attractors, it will be a neural network for classification.

On the other hand, there are two major differences from ordinary deep neural networks. First, J_{ij} is symmetric with respect to the indices, and the same $J_{i,j}$ is shared in all the layers. In this sense, it can be called a sparse network. Second, the way of training is completely different. These differences can be thought of as differences in learning methods depending on the purpose.

The Hopfield model we saw in this chapter is a primitive model of a neural network that has historically led to Boltzmann machines (see Chap. 6), and we should recognize that the model is deeply related to typical properties of spinglass systems. When discussing the relationship between physics and machine learning, it is important not only to know what you want to train, but also to physically interpret the neural network itself. In condensed matter physics that deals with manybody systems in physics, spin models are fundamental, and various spin models dominate intriguing physical phenomena and phase diagrams. From this point of view, the Hopfield model similar to a spinglass can be a starting point in the future development of physics and machine learning.

Chapter 11
Quantum Manybody Systems, Tensor Networks and Neural Networks

Abstract In condensed matter physics, finding the wave function of a quantum many-body system is the most important issue. Theoretical development in recent years includes a wave function approximation using a tensor network. At first glance, the tensor network looks very similar to neural network diagrams, but how are they actually related? In this chapter we will see the relation and the mapping, and that the restricted Boltzmann machine is closely related to tensor networks.

Microscopically, physics is a quantum system, and all isolated quantum systems are governed by wave functions. For example, for simplicity, let us consider a situation with zero temperature. Given a Hamiltonian of a system, the quantum phase is determined by its lowest energy state (the ground state). To obtain a concrete wave function is one of the most important ways to characterize quantum systems.

Hilbert space, which builds states of quantum many-body systems, suffers from enormous combinatorial possibilities. For example, consider a system with N **qubits**. A qubit is a system with two states, $|0\rangle$ and $|1\rangle$, and is quantum mechanically the same as a system with a spin $\hbar/2$. In this case, the Hilbert space has 2^N dimensions, so the number of states increases exponentially with the number of degrees of freedom. Since the problem of finding the ground state of a quantum many-body system is the task of selecting only one state from this huge Hilbert space, some "physical sense" is needed. That is where machine learning approaches come into play.

Until now, methods have been developed that minimize the energy in a subspace, considering only the states described by a relatively small number of parameters in the Hilbert space. A method of constructing a subspace particularly efficiently is called a tensor network.[1] There are various types of tensor networks depending on their physical meaning and their targets. On the other hand, in this book we have described various neural networks and how to use them. According to the neural network universal approximation theorem (see Chap. 3), any function can

[1]The tensor network is not a physical network arranged spatially. It is a graph specifying how spins etc. are intertwined with each other.

A. Tanaka et al., *Deep Learning and Physics*, Mathematical Physics Studies,
https://doi.org/10.1007/978-981-33-6108-9_11

be approximated by increasing the number of units sufficiently. Therefore, we can use neural networks as a method of constructing quantum states.

In the following, we will introduce the methods proposed in literature [122] and the results, and also look at the differences between tensor networks and neural networks.

11.1 Neural Network Wave Function

First, let us consider how the **wave function** of a quantum system can be represented by a neural network. A wave function of an N-qubit system is the coefficient $\psi(s_1, \cdots, s_N)$ in the state $|\psi\rangle$,

$$|\psi\rangle = \sum_{s_1,\cdots,s_N} \psi(s_1, \cdots, s_N)|s_1\rangle \cdots |s_N\rangle \,. \tag{11.1}$$

Here, $s_a = 0, 1$ is the basis of the qubit, and ψ is a complex function using it as a variable. That is, the nonlinear function ψ transforms the input such as $(s_1, \cdots, s_N) = (0, 0, 1, 0, 1, \cdots)$ into a complex-valued output $\psi(0, 0, 1, 0, 1, \cdots)$. The nonlinear function ψ defines the quantum state. When this nonlinear function is determined so that the energy of the system

$$E = \frac{\langle\psi|H|\psi\rangle}{\langle\psi|\psi\rangle} \tag{11.2}$$

is minimized, we call the state a **ground state**.

Therefore, if we recapture this problem as machine learning, we represent the nonlinear function ψ by a neural network and consider the error function as the system energy E. Hence, this is not supervised learning. In supervised machine learning, the error function is defined as the difference from the correct output, because there is a correct combination of an input and an output, and training is performed so that a function provides the correct output when the input is given. On the other hand, in the problem of finding the ground-state wave function, the energy of the system is adopted as the error function, and the wave function is trained so that the energy becomes small.

Now, according to the literature [122], let us express the wave function by a **restricted Boltzmann machine.**[2] As seen in Chap. 6, the restricted Boltzmann machine provides a Boltzmann weight factor as an output. The relation between the hidden layer unit and the input unit is given with a spin Hamiltonian. If the unit

[2]See also [123]. A physical interpretation of the neural network is described in [124].

value in the hidden layer is $h_1, \cdots, h_M$, the output will be as follows:

$$\psi(s_1, \cdots, s_N) = \sum_{h_A} \exp\left[\sum_a a_a s_a + \sum_A b_A h_A + \sum_{a,A} J_{aA} s_a h_A\right]. \tag{11.3}$$

The exponent is the "**Hamiltonian**" to produce the Boltzmann factor. Note that this Hamiltonian has nothing to do with the Hamiltonian of the quantum system we are thinking about. a_a and b_A are biases, and J_{aA} is the weight of the neural network. One can explicitly make the summation $\sum_{h_A}$, to find

$$\psi(s_1, \cdots, s_N) = e^{\sum_a a_a s_a} \prod_{A=1}^{M} 2\cosh\left[b_A + \sum_a J_{aA} s_a\right]. \tag{11.4}$$

Based on this expression, one updates the bias a_a, b_A and the weight J_{aA}, and lowers the energy (11.2) of the system. The update procedure is the same as for the previous supervised learning.

Let us look at the results of an actual application example. Consider a two-dimensional Heisenberg model with an antiferromagnetic Hamiltonian. The Hamiltonian uses the spin operators $\hat{\sigma}^x, \hat{\sigma}^y, \hat{\sigma}^z$,

$$\mathcal{H} = \sum_{\langle a,b\rangle} \left[\hat{\sigma}_a^x \hat{\sigma}_b^x + \hat{\sigma}_a^y \hat{\sigma}_b^y + \hat{\sigma}_a^z \hat{\sigma}_b^z\right]. \tag{11.5}$$

Here $\langle a, b\rangle$ represents two adjacent lattice points on a 10×10 periodic square lattice. In this system, after the weight and bias in the wave function using a restricted Boltzmann machine are updated, a low-energy wave function is obtained. According to the literature [122], as shown in Fig. 11.1, the result is that the neural network wave functions give less energy than an energy value ("EPS" or "PEPS") obtained

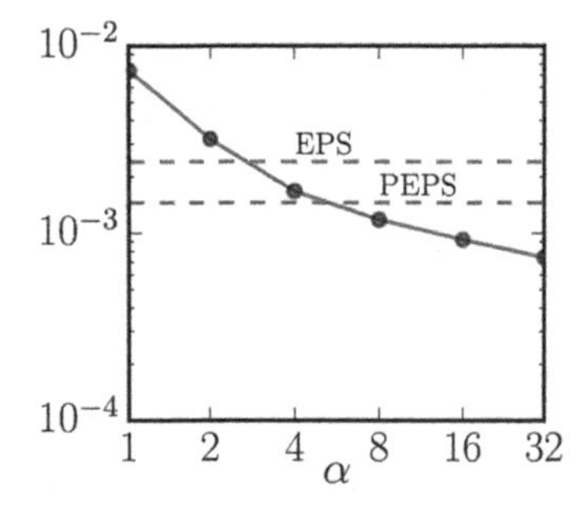

Fig. 11.1 Energy of the two-dimensional antiferromagnetic Heisenberg model, given in [122]. EPS and PEPS are energy minimizations using wave functions written by conventional tensor networks. α represents the number of hidden units. The more hidden units, the lower the energy, indicating that we are approaching the true ground state

by optimizing the wave function of the conventional method. And it can be seen that the more hidden units α, the better the neural network wave function is.

11.2 Tensor Networks and Neural Networks

Tensor networks provide a way to efficiently represent wave functions of quantum many-body systems, that is, a way to construct subspaces in Hilbert space. So let us discuss the relationship with neural networks here.

11.2.1 Tensor Network

The state represented by the simplest tensor network is the case with $N = 2$,

$$\sum_{i,j=0,1} A_{ij}|i\rangle|j\rangle \,. \tag{11.6}$$

Here, $i, j = 0, 1$ is an index for the spin, and A_{ij} is called a **tensor**. The tensor A has $2 \times 2 = 4$ components, so it has four complex degrees of freedom. In the following, we will use Einstein's convention[3] and omit the $\sum$ symbol. Next, consider the following state of $N = 4$:

$$B_{mn}A_{mij}A_{nkl}|i\rangle|j\rangle|k\rangle|l\rangle \,. \tag{11.7}$$

Tensor A has three indices and tensor B has two indices. They are contracted by indices m, n. In the $N = 2$ example (11.6), the tensor parameter A can represent all of the Hilbert space, but in the $N = 4$ example (11.7), it is not possible. This is because if $N = 4$, the entire Hilbert space is spanned by $2^4 = 16$ complex numbers, but the degree of freedom of the tensor is $2^3 = 8$ for A and $2^2 = 4$ for B, thus only 12 in total. In other words, a subspace of the Hilbert space is parameterized.

It is standard to use a graph to represent the wave function by the tensor A or B. The typical notation is that a tensor has legs whose number is that of the indices (see Fig. 11.2). It appears to be similar to the neural network notation, but note that the meaning of the line is completely different in the following ways: In a tensor network, tensors are represented by squares, triangles, and circles. The line (leg) extending from it has the meaning of subscript $i = 0, 1$, and so, the line means an input or an output. Since a tensor with three indices has three legs, some of the three are inputs and the remaining lines are outputs. In neural networks, on

[3]Einstein's convention is the understanding that indices that appear more than once will be summed over.

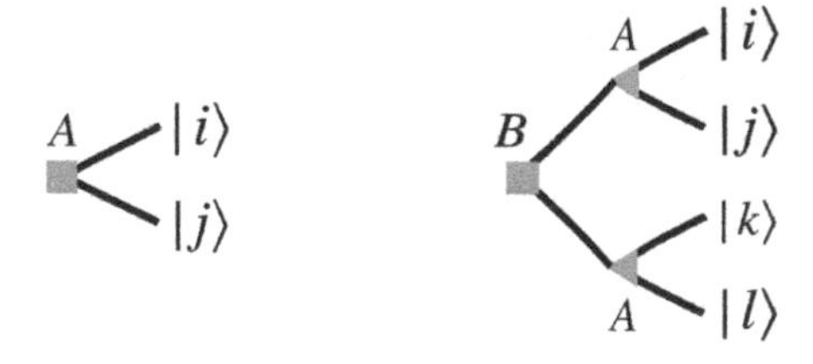

Fig. 11.2 Quantum states (11.6) and (11.7) represented by tensor networks

the other hand, the lines mean "multiplying weights W," which is the role of a tensor in a tensor network. Also, in the neural network, the units (circles) have the meaning of the input and the output, and in the tensor network, circles are tensors (the role of weighting). Therefore, the meaning of lines and circles is reversed in tensor networks and neural networks.

11.2.2 Tensor Network Representation of Restricted Boltzmann Machines

Now let us look at the relationship between the restricted Boltzmann machines and the tensor networks [125]. In the neural network graph that gives a restricted Boltzmann machine, we associate the following 2×2 matrix with each line connecting a visible unit s_a and a hidden unit h_A,

$$M^{(aA)}_{ss'} \equiv \begin{pmatrix} 1 & 1 \\ 1 & \exp[J_{aA}] \end{pmatrix}_{ss'}, \tag{11.8}$$

and also associate the following 2×2 matrix at the connection of the unit and the line,

$$\Lambda^{(a)}_{ss'} \equiv \begin{pmatrix} 1 & 0 \\ 0 & \exp[a_a] \end{pmatrix}_{ss'}, \quad \tilde{\Lambda}^{(A)}_{ss'} \equiv \begin{pmatrix} 1 & 0 \\ 0 & \exp[b_A] \end{pmatrix}_{ss'}. \tag{11.9}$$

With this rule, all restricted Boltzmann machines can be interpreted as tensor networks.

For example, consider the simplest restricted Boltzmann machine made of a single visible unit, a single hidden unit and a single line connecting them (Fig. 11.3 left). According to the above rules,

$$\tilde{\Lambda}^{(a=1)} M^{(1,1)} \Lambda^{(A=1)} = \begin{pmatrix} 1 & e^{a_1} \\ e^{b_1} & e^{a_1+b_1+J_{11}} \end{pmatrix}. \tag{11.10}$$

The elements of this matrix correspond to the weights of the restricted Boltzmann machine to which $(s, h) = (0, 0), (0, 1), (1, 0), (1, 1)$ is substituted. To make a

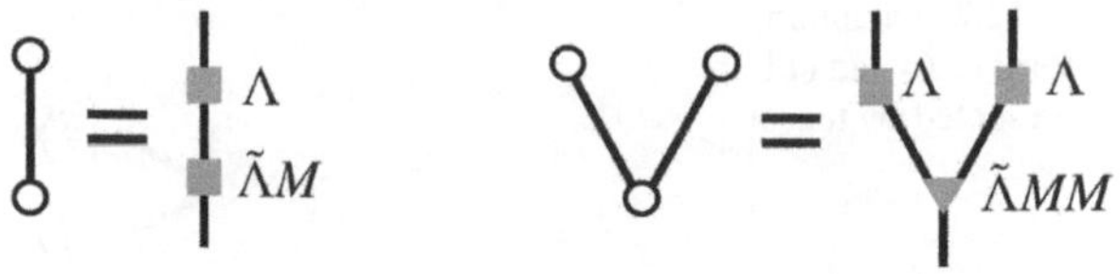

Fig. 11.3 Tensor network representation of a restricted Boltzmann machine

summation over hidden unit variables, we just need to take a product with the vector (1, 1),

$$\psi(s) = \left[(1,1)\,\tilde{\Lambda}^{(1)}M^{(1,1)}\Lambda^{(1)}\right]_s \tag{11.11}$$

which is equivalent to (11.3). The expression (11.10) describes the restricted Boltzmann machine as a product of matrices, and since the elements of the final matrix product represent the components of the unit, it is exactly a tensor network representation.

Let us consider a slightly more complicated example: a restricted Boltzmann machine as shown in the right panel of Fig. 11.3. According to the rules, we have a matrix product

$$\tilde{\Lambda}^{(A=1)}_{st} M^{(a=1,A=1)}_{tt'} M^{(a=2,A=1)}_{tt''} \Lambda^{(a=1)}_{t's'} \Lambda^{(a=2)}_{t''s''} \,. \tag{11.12}$$

Here the matrix M is multiplied twice as the number of lines is two, and the matrix $\Lambda, \tilde{\Lambda}$ is multiplied three times as the number of units is three. Note that the index t appears three times, and the above expression omits $\sum_t$. The sum over an index that appears three times cannot be expressed by a matrix multiplication. Therefore, we define

$$\tilde{M}_{st't''} \equiv \tilde{\Lambda}^{(A=1)}_{st} M^{(a=1,A=1)}_{tt'} M^{(a=2,A=1)}_{tt''} \,, \tag{11.13}$$

which works well and will be useful later. The new entry $\tilde{M}$ has three indices and corresponds to a three-legged tensor. So again, we found that this restricted Boltzmann machine has a representation in a tensor network. Looking at the right panel of Fig. 11.3, a tensor network representation is given.

Thus, the restricted Boltzmann machine allows tensor expressions. On the other hand, is a general tensor network represented by a neural network? In fact, it has been proven that every state composed of tensor networks (that is, every state that can be represented by a quantum circuit) can be represented by a deep Boltzmann machine [126].[4] Therefore, tensor networks often used in condensed matter physics are well suited for machine learning in the sense of Boltzmann machines. From this perspective, various studies are underway.

[4]The relations among the parameters of the networks are discussed in [127].

What about a feedforward neural network instead of a Boltzmann machine? As we saw in Chap. 3, feedforward neural networks have a universal approximation theorem, so any nonlinear function can be expressed, and in this sense any tensor network can be expressed. However, constructing a neural network to represent a given tensor network has some difficulties, as we describe below.

First, in the case of Fig. 11.3 right, two A's are connected to the two legs of tensor B, and the state of the tensor network includes a multiplication of A. In general, tensors combine by multiplication, which creates nonlinearities, whereas feedforward neural networks do not, and create nonlinearities in the form of activation functions acting on a unit-by-unit basis. To make up for this difference, you need to use an activation function in the form of a multiplication.[5]

A tensor network that can be expressed as a neural network using such a new activation function is a tree graph. A tensor network of graphs with loops inside (which also includes what is called MERA[6]), instead of a tree, is difficult to express with a feedforward type. This is because the loop graph has two or more outputs in the middle layer, and tensors connected to them appear in parallel (the tensor product structure), but in a feedforward neural network, when tensors arranged in parallel are recombined, it allows only one of them to be expressed with priority, and the parallelism of the tensor is hindered.

As described above, tensor networks are different from neural networks in many ways. However, each has its advantages and can be moved back and forth through common concepts. As explained in Chap. 10, neural networks originate from physical many-body systems such as Hopfield models, so they are compatible with physical systems. In the future, interconnection of quantum many-body systems and machine learning will advance in various forms.

[5] There is also a study that uses product pooling to see a correspondence [128].

[6] MERA stands for Multiscale Entanglement Renormalization Ansatz.

Chapter 12
Application to Superstring Theory

Abstract The last chapter describes an example of solving the inverse problem of string theory as an application of deep learning. The superstring theory unifies gravity and other forces, and in recent years, the "holographic principle," that the world governed by gravity is equivalent to the world of other forces, has been actively studied. We will solve the inverse problem of the emergence of the gravitational world by applying the correspondence to the dynamical system seen in Chap. 9, and look at the new relationship between machine learning and spacetime.

Superstring theory has been studied as quantum theory of gravity and as a theory that can describe all the forces in the universe in a unified way.[1] This chapter shows examples of applying machine learning, especially deep learning, to mathematical problems in string theory. Despite its history, a variety of studies that apply machine learning techniques to physics in earnest has only recently opened up. The content of this chapter is mainly based on collaborative research between the authors and Sotaro Sugishita [118, 130], and other kinds of research on string theory have been done, so diverse progress is expected.

First, let us outline two of the inverse problems in string theory from a general perspective without using mathematical formulas. Next, we explain how to regard deep neural networks as spacetimes, as one of the methods to solve an **inverse problem** in the holographic principle.

12.1 Inverse Problems in String Theory

There are two major characteristics of the mathematical achievements of superstring theory that enable the quantization of gravity. The first is the constraint that the spacetime needs to be 10-dimensional, and the second is that it results in various gauge symmetries and sets of elementary particles which transform under them.

---

[1]For an introductory book on string theory, see [129].

A. Tanaka et al., *Deep Learning and Physics*, Mathematical Physics Studies,
https://doi.org/10.1007/978-981-33-6108-9_12

From these two characteristics, it is not an exaggeration to say that research on string theory has greatly advanced. In the following, we will mention that there are two important inverse problems in string theory. The first inverse problem is the problem of compaction, and the second inverse problem is the holographic principle.[2]

12.1.1 Compactification as an Inverse Problem

The first constraint and the second point are related. The spacetime dimension of our universe is 4, so if it was originally a 10-dimensional spacetime, the 6-dimensional space must be a very small and compact space. If we adopt what is generally called a Calabi–Yau manifold as the space, it is known that the resulting 4-dimensional spacetime theory will be closer to our familiar elementary particle theory. There are many types of Calabi–Yau manifolds, all of which are mathematically allowed in perturbative string theory. If one Calabi–Yau manifold is adopted as a compact 6-dimensional internal space, the types, numbers, and symmetry of the remaining 4-dimensional spacetime elementary particles will be determined accordingly.

In this sense, string theory leaves various possibilities. Instead of deriving a single elementary particle model in 4-dimensional spacetime, it derives many types of elementary particle models. Since there are a finite number of Calabi-Yau manifolds, mathematically the number of models is finite, but the number is very large. Of course, there is an infinite variety of types of quantum field theory that do not include gravity, and string theory imposes restrictions on quantum field theories in that sense. However, the restrictions are not very strong, and various elementary particle models are allowed.[3]

Therefore, if we consider string theory as a theory describing our universe, we need to find a Calabi–Yau manifold that leads to a 4-dimensional model closer to the standard model of elementary particles. So far, a number of Calabi-Yau varieties are known that provide particle content and symmetry close to that of the standard model of elementary particles. However, the research is not complete.

In addition, when deriving a 4-dimensional elementary particle model from string theory, it is not only a matter of choosing the Calabi–Yau manifold. For example, D-branes[4] on which strings can end are introduced into string theory. D-branes of various dimensions can be considered, and by placing them in 10-dimensional

[2]These are mutually related, but it is too technical, so here we will just give an overview of each.

[3]Of course, further progress in string theory research may prove even more restrictive. There are two reasons for this. First, the region where the superstring theory as the quantum theory of gravity is well understood is where the perturbational picture holds, that is, when the coupling constant of the string is small. So, once the non-perturbative properties is understood, stronger restrictions could be imposed. Second, quantum gravity theory is not always superstring theory. In the AdS/CFT correspondence which will be described later, quantum gravity theory is defined more widely, and in that sense the string theory is extended.

[4]See [129] for an introduction to D-branes.

spacetime, various symmetries can be realized. By placing a good combination of D-branes, it is possible to reproduce the symmetry of the standard model of elementary particles and the particle content. Creating a 4-dimensional spacetime elementary particle model (or quantum field theory) in this way by combining D-branes in string theory is generally called brane construction. Various methods have been combined to create a 4-dimensional spacetime elementary particle model.

At present, a method called F-theory compactification has been developed, in which compaction by Calabi–Yau manifolds from 10 dimensions and combinatrics of the D-branes are approximately integrated. The F-theory is defined as a 12-dimensional spacetime, and its 2-dimensional compaction reproduces the type IIB string theory. When the 8-dimensional compaction is performed, an elementary particle model is created in the remaining 4-dimensional spacetime. This 8-dimensional compact space employs a complex 4-dimensional Calabi–Yau manifold. From this standpoint, if the type of complex 4-dimensional Calabi–Yau manifold is determined, the elementary particle model of 4-dimensional spacetime is determined, so the type of Calabi–Yau variety needs to be scrutinized.

As you can see, the problem of deriving the standard model of elementary particles from string theory is an inverse problem. The problem is to select a manifold necessary for compaction of string theory so that it becomes the standard model of elementary particles. One of the difficulties with this problem is that the Calabi-Yau manifold is not well understood (for example, it is difficult to even know the metric on the manifold). For exploration of compact space, machine learning that is effective for inverse problems has begun to be used [131–134].

12.1.2 The Holographic Principle as an Inverse Problem

The progress of string theory research for the past 20 years has been carried by the **AdS/CFT correspondence**. The AdS/CFT correspondence discovered by J. Maldacena in 1997 [135–137] is regarded as a concrete realization of the **holographic principle**, originally proposed by G. 'tHooft and generalized by L. Susskind [138].

The holographic principle is that a quantum theory including gravity is equivalent to a quantum field theory without gravity defined in a lower-dimensional spacetime. The first example given by Mardacena is the equivalence between type IIB superstring theory on an $AdS_5 \times S^5$, that is, a 5-dimensional **Anti-de Sitter spacetime** and a 5-dimensional sphere, and the $\mathcal{N} = 4$ supersymmetric Yang–Mills theory on a flat 4-dimensional spacetime. It was suggested by considering a special limit in the 10-dimensional string theory with D-branes. It is the equivalence of these two theories: the former is a quantum gravity theory, while the latter is a lower-dimensional quantum field theory without gravity. It is a concrete example of the holographic principle. Generally, the former is called "gravity side" or "AdS side," and the latter is called the "CFT side" or "boundary side" or "gauge theory side." CFT is an abbreviation of conformal field theory, which is a field theory that is invariant under conformal transformation. The $\mathcal{N} = 4$ supersymmetric Yang–Mills theory is an example of CFT.

In the first example by Maldacena, the gravity side is a quantum gravity theory (i.e., superstring theory), but the classical limit and low energy region are often considered. This is because in that area, the quantum gravity theory becomes a classical Einstein theory (in higher-dimensional spacetimes with some supersymmetries), and Lagrangians are well known and easy to handle. It is known that taking such a limit on the gravity side is equivalent to taking the following two limits in the Yang–Mills theory on the CFT side. First you need the large N limit, where N is the gauge symmetry $SU(N)$ of the Yang-Mills theory. By taking N to infinity, the gravity side becomes classical (quantum effects can be ignored). Next, the strong coupling limit is needed. This is called the large λ limit, where $\lambda \equiv Ng^2$ is called the 't Hooft coupling constant, and g is the coupling constant of the Yang–Mills theory. By making λ infinite, on the gravitational side, the higher-order derivatives acting on the fields can be ignored.

There are two important open questions of the holographic principle. The first is to give a proof of the example which Maldacena originally proposed. The second is to give a condition on what quantum field theory on the boundary side can allow a gravitational description. These problems are, of course, thought to be related to each other, but the first example of Maldacena was suggested by some limit of string theory, while the second problem is recognized as an issue that needs to be discussed beyond superstring theory. In particular, as the second problem is elucidated, we will gain important insights on whether string theory is the only method for quantum theory of gravity. The difficulty in solving these problems is that, as mentioned above, the Yang–Mills theory, which is the gauge theory side, has a strong coupling, so that it is not possible to use the usual perturbation method, so the research approach is difficult.

Against this background, various examples of the AdS/CFT correspondence have been built. The AdS/CFT correspondence is generally called "gauge/gravity duality." The main early examples were derived from string theory, by deforming the examples proposed by Maldacena. Furthermore, many examples have been published that go beyond the derivation from string theory. Those examples have properties that are expected to be common to the gauge and gravity sides (such as symmetries).

In the standard model of elementary particles known to us, the part responsible for the strong force is written by **quantum chromodynamics** (QCD), which is based on $SU(3)$ Yang–Mills theory. It is known that various hadron pictures appear at low energies due to its strong coupling. Solving QCD is important for indicating possible deviation of the experimental data from the standard model. Moreover, as the best studied example of quantum theory of strongly coupled fields, it has great mathematical significance and many researchers are attracted. There is a research field that applies the gauge/gravity correspondence to this QCD and performs various calculations on the gravity side to evaluate hadron physical quantities. This field is called **holographic QCD**.

Researchers are experimenting with various gravity models because no gravity-side theory equivalent to QCD has been found. On the gravity side, one can write a spacetime metric in a spacetime with one dimension higher, and a Lagrangian of various matter fields in it. One can perform various classical calculations based on that, and can use the dictionary for the gauge/gravity duality to calculate the QCD physical quantities which live on the boundary side. So what is the gravity theory equivalent to QCD?

As you can see, this is one of the inverse problems. QCD is a well-defined quantum field theory whose Lagrangian is given, and its physical quantities have been widely studied. Because of the strong coupling, it is not easy to calculate, but it has been established that it is possible to evaluate various physical quantities by numerical calculation using a supercomputer: lattice QCD.[5] In particular, as for the spectrum (mass distribution) of hadrons, which are bound states of elementary quarks in QCD, the calculation results of the lattice QCD match very well with the measurement results of accelerator experiments, thus our world is confirmed to be described by QCD. The problem of finding a gravitational description of such a given quantum field theory is an inverse problem. This is because, until now, almost all holographic QCD calculations have been performed for a given gravitational-side Lagrangian by hand and are interpreted as QCD-like quantities.

In the next section, we will introduce a method using deep learning [118, 130] to solve this inverse problem. In this method, the data of the lattice QCD simulation can be used to determine the metric on the gravity side so that it reproduces the data, by learning.

One of the fundamental problems of the gauge/gravity correspondence is how to construct the gravity side for a given quantum field theory on the boundary side, and how to judge whether there is a description of the gravity side in the first place. This kind of problem is not limited to QCD. In recent years, wider gauge/gravity correspondences based on entanglement entropy and quantum information theory have been studied. The concept of "emergent gravity" is gaining widespread acceptance and there is hope for how the inverse problem will be solved in the future.

12.2 Curved Spacetime Is a Neural Network

In order to partially solve the inverse problem of finding the gravitational theory emerging from the boundary quantum field theory such as QCD, the following simple settings are made. We shall regard the gravitational spacetime itself as a neural network.

---

[5]This is the field of research where the inside of the nucleus can be numerically calculated at the quark/gluon level from the first principle. Due to the large amount of numerical calculation, calculations are performed by the Markov chain Monte Carlo method using a supercomputer. The formulation itself is analogous to statistical mechanics. For details, see [104].

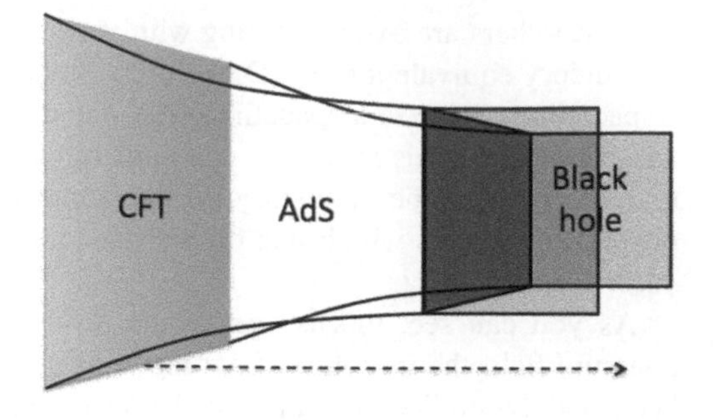

Fig. 12.1 Conceptual diagram for AdS/CFT. This figure and the conceptual diagram of a typical deep neural network (Fig. 9.1) are considered the same, for the inverse problem to be solved

Given the gravity theory, to find the one-point function of quantum field theory on the boundary side, we just simply solve the differential equation on the gravity spacetime (called "bulk"). In particular, if the gravity side is classical, we solve classical differential equations. Information about the one-point function of the quantum field theory of the boundary side is provided in the value (at the region near the boundary) of the solution of the differential equation on the gravity side. Therefore, determining the metric on the gravity side from the quantum field theory on the boundary side means the following question: When the boundary value of a field is given, what is the differential equation that reproduces it as a solution? This is an example of an inverse problem, as explained in Chap. 7.

At this point, we consider that the theory of gravity itself is regarded as a neural network, and the weight of the neural network is regarded as the metric of gravity, so that the metric is read from the trained weights. For the training data, we use a one-point function of quantum field theory of the boundary side. Since this gives the value of the field at the boundary in gravity theory, it just functions as the input data of the neural network.

12.2.1 *Neural Network Representation of Field Theory in Curved Spacetime*

Let us rewrite gravity theory with a neural network. The meaning of this is clear when you look at Figs. 9.1 and 12.1. To build concrete relationships, neural network expressions of various physical systems introduced in Chap. 9 are useful. In particular, the form of the time evolution of the Hamiltonian system can be written as a neural network, so the gravity theory can be written in the same way as a neural network by regarding the emergent spatial direction in the holographic principle as the time direction of the Hamiltonian system. Specifically, a neural network is a nonlinear relation between input data $x^{(1)}$ and output data y, given by

$$y(x^{(1)}) = f_i \varphi(J_{ij}^{(N-1)} \varphi(J_{jk}^{(N-2)} \cdots \varphi(J_{lm}^{(1)} x_m^{(1)}))) \,. \tag{12.1}$$

Here J is the weight and φ is the activation function.

First, let us consider a curved spacetime of $(d + 1)$ dimensions. We consider the theory of gravity on that, but here, for simplicity, consider instead a theory of a scalar field ϕ on a curved spacetime. The scalar field theory action is given as follows:

$$S = \int d^{d+1}x\sqrt{-\det g}\left[-\frac{1}{2}(\partial_\mu\phi)^2 - \frac{1}{2}m^2\phi^2 - V(\phi)\right]. \tag{12.2}$$

Let the spacetime of the quantum field theory on the boundary side be flat and d-dimensional, and let the direction of the space emerging on the gravity side be η. The metric of the curved spacetime of the gravity side can be generally written as

$$ds^2 = -f(\eta)dt^2 + d\eta^2 + g(\eta)(dx_1^2 + \cdots + dx_{d-1}^2). \tag{12.3}$$

Here $f(\eta)$ and $g(\eta)$ are metric components and should be determined by learning. In our gauge the metric component in the η direction is equal to 1. At this point, $f(\eta)$ and $g(\eta)$ are undetermined, but the following two conditions must be satisfied in order for AdS/CFT correspondence to work. First, at the boundary $\eta \to \infty$, the spacetime needs to be asymptotically AdS, $f \simeq g \simeq \exp[2\eta/R]$ $(\eta \sim \infty)$, where R is the AdS radius. The other condition is for the opposite boundary of the curved spacetime. Assuming that the quantum field theory on the boundary side is a finite temperature system, as a result, the gravity side becomes a **black hole** spacetime. So, the other boundary condition is the event horizon of the black hole. This is expressed as $f \propto \eta^2$, $g \simeq \text{const}$ $(\eta \sim 0)$.

From this action (12.3), the **equation of motion** of the field $\phi(\eta)$ is an ordinary differential equation

$$\partial_\eta\pi + h(\eta)\pi - m^2\phi - \frac{\delta V[\phi]}{\delta\phi} = 0, \qquad \pi \equiv \partial_\eta\phi. \tag{12.4}$$

Here, as explained in Chap. 9, we introduced $\pi(\eta)$ as the conjugate momentum for ϕ, and transformed the differential equation into a set of the first-order differential equations. The metric components are included as $h(\eta) \equiv \partial_\eta \log\sqrt{f(\eta)g(\eta)^{d-1}}$. In addition, we discretize the η direction as we did in Chap. 9,

$$\phi(\eta + \Delta\eta) = \phi(\eta) + \Delta\eta\,\pi(\eta), \tag{12.5}$$

$$\pi(\eta + \Delta\eta) = \pi(\eta) - \Delta\eta\left(h(\eta)\pi(\eta) - m^2\phi(\eta) - \frac{\delta V(\phi)}{\delta\phi(\eta)}\right). \tag{12.6}$$

This is a neural network representation of the equation of motion for a scalar field ϕ in a bulk curved spacetime. A schematic picture is shown in Fig. 12.2. In particular,

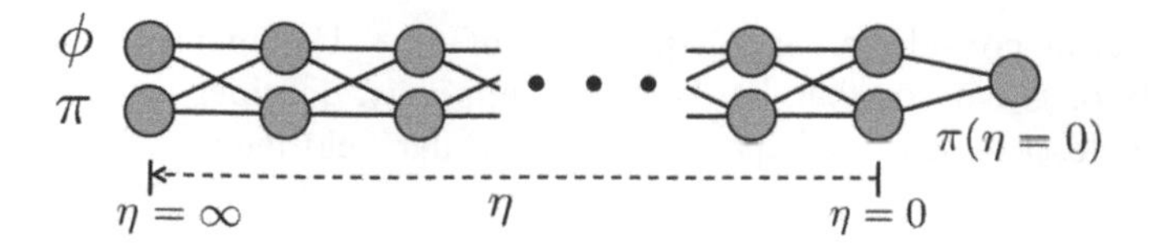

Fig. 12.2 A neural network representation of the equation of motion of a scalar field on a curved spacetime

since the η direction is discretized as $\eta^{(n)} \equiv (N - n + 1)\Delta\eta$, the input data of the neural network is $(\phi(\infty), \pi(\infty))^{\mathrm{T}}$, and the neural network weight J is

$$J^{(n)} = \begin{pmatrix} 1 & \Delta\eta \\ \Delta\eta\, m^2 & 1 - \Delta\eta\, h(\eta^{(n)}) \end{pmatrix} . \tag{12.7}$$

So it is interpreted to include the metric information as a part of it. Note that some weights are fixed to be 1. Namely, this is a sparse network. The activation function is chosen as follows to reproduce (12.6):

$$\begin{cases} \varphi(x_1) = x_1 \, , \\ \varphi(x_2) = x_2 + \Delta\eta \, \dfrac{\delta V(x_1)}{\delta x_1} \, . \end{cases} \tag{12.8}$$

In this way, using the simplest deep neural network (Fig. 12.2), if the weights and the activation functions are chosen as (12.7) and (12.8), the classical equation of motion of the scalar field in the gravity side can be expressed as a neural network (12.1).

12.2.2 How to Choose Input/Output Data

If the bulk differential equation is regarded as a neural network in this way, the boundary conditions of the differential equation automatically become the input data and the output data. In the case of AdS/CFT correspondence, it is known that the behavior of the solution of the differential equation in the asymptotic AdS region ($\eta \to \infty$) corresponds to a one-point function of the quantum field theory of the boundary side. So you can choose that one-point function as the input data. On the other hand, the boundary condition on the other boundary ($\eta = 0$) of the bulk spacetime will be the boundary condition imposed by the event horizon of the black hole. At this stage, we can ask: what is the training? The training is to adjust the weight of the neural network, that is, the gravity metric, so that when the "correct" data is input from the $\eta \to \infty$ boundary, the black hole horizon boundary condition is correctly satisfied at $\eta = 0$. This is a function-degree-of-freedom optimization problem, so, it is a machine learning.

The relation between the boundary value of the scalar field in the asymptotic AdS region and the one-point function of the quantum field theory on the boundary side is specified by the general principle defined in the literature [139]. Suppose that the operator of quantum field theory on the boundary side is O, and the term gO exists in the Lagrangian of the quantum field theory as its source term. Given the source g, the one-point function $\langle O \rangle$ means that the vacuum expectation value $\langle O \rangle$ is given as a function of g. For example, when an external magnetic field is given as a source, $\langle O \rangle$ is a magnetization operator, that is, the response to the external field. In QCD, for example, if the source g is the mass of the quark, the mass term of the QCD is $m_q \bar{\psi}\psi$, so the chiral condensate $\langle \bar{\psi}\psi \rangle$ corresponds to the response. In this manner, when g and $\langle O \rangle$ are obtained by the quantum field theory on the boundary side, the AdS/CFT correspondence tells us that there is the following relation [139][6] on the gravity side, under the normalization condition that the AdS radius R is 1,

$$\phi(\eta_{\text{ini}}) = g \exp[-\Delta_- \eta_{\text{ini}}] + \langle O \rangle \frac{\exp[-\Delta_+ \eta_{\text{ini}}]}{\Delta_+ - \Delta_-} . \tag{12.9}$$

Here Δ is determined from the mass of the bulk, as $\Delta_\pm \equiv (d/2) \pm \sqrt{d^2/4 + m^2 R^2}$, and Δ_+ is the **conformal dimension**[7] of the operator O. In addition, a cutoff was made in the region $\eta \to \infty$: the boundary is given at $\eta = \eta_{\text{ini}}$. Differentiating this relation by η gives

$$\pi(\eta_{\text{ini}}) = -g\Delta_- \exp[-\Delta_- \eta_{\text{ini}}] - \langle O \rangle \frac{\Delta_+ \exp[-\Delta_+ \eta_{\text{ini}}]}{\Delta_+ - \Delta_-} . \tag{12.10}$$

These are the formulas that give the values of ϕ and π at $\eta \to \infty$ for a given one-point function of the quantum field theory.

On the other hand, what about the output data? At the even horizon of the black hole, only waves that enter the horizon are allowed. Therefore, the solution of the equation of motion only allows wave solutions in the negative direction of η. Specifically, we can find out (by restoring the time derivative term as well) the following condition:

$$0 = F \equiv \left[\frac{2}{\eta}\pi - m^2\phi - \frac{\delta V(\phi)}{\delta \phi} \right]_{\eta = \eta_{\text{fin}}} . \tag{12.11}$$

[6]When the field is evaluated by its integration in the bulk coordinates, the first term is generally a non-normalizable mode, and the second term is a normalizable mode.

[7]The conformal dimension Δ means that is in quantum field theory which is invariant under conformal symmetry, the operator O transforms, when the coordinate scales as $x \to \alpha x$, like $O \to \alpha^{-\Delta} O$. In a free field theory, Δ is the mass dimension of the operator (or the field).

Here $\eta = \eta_{\rm fin} \sim 0$ is the cutoff near the horizon. In $\eta \to 0$, only the first term of F is important, so simply the horizon condition is as follows:

$$\pi = 0\,. \tag{12.12}$$

That is, whether the input data is correct or not is determined by whether the value of π is zero at the last layer of the neural network.

Of course, since this is a numerical experiment, training will not progress at all if you request that the output π be completely zero. Therefore, a trick is needed to judge that the data is correct when the value is close to 0. This can be done by setting a threshold: for example, if the value of the magnitude of the output is less than 0.1, it is considered to be 0. Also, the value of π in the final layer may be very large, while considering that neural networks are good at binary classification problems, it is also effective to restrict the value range by converting the value of π in the final layer to $\tanh(\pi)$. With that modification, if "incorrect data" that does not reproduce the one-point function is input, the training is made such that the output should be 1 instead of 0.

In this way, we have a method to solve the inverse problem of determining the gravity metric from the quantum data of the boundary quantum field theory. The important thing in this "transition" of the problem was not only to rewrite the differential equation into a neural network form, but also to divide the differential equation and the boundary value problem into unknown and given, so that the unknown is the network weights, and the known data is the input and the output. If such separation is possible, the inverse problem can be solved using the techniques described in this chapter.

12.3 Emergent Spacetime on Neural Networks

Now, let us actually implement the above neural network on a computer and perform a numerical experiment to see if spacetime metrics can be obtained by learning. Here, we introduce two independent numerical experiments [118]. The first one is a numerical experiment to see if a known metric can be reproduced. And the second one is to answer the question: when a neural network is fed with experimental data of a real material whose gravity side is not known at all, will a spacetime emerge?[8] **Time and space emerging from QCD data** [130] will be described in more detail in the next section.

[8] In physics, the term "emergent" refers to a phenomenon in which properties and equations that are not expected from the degrees of freedom that define a physical system appear dynamically.

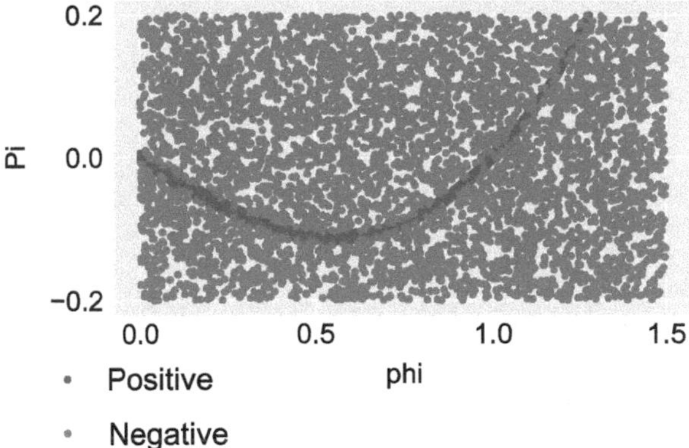

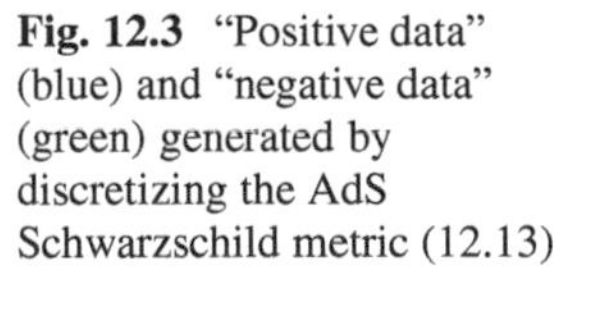
Fig. 12.3 "Positive data" (blue) and "negative data" (green) generated by discretizing the AdS Schwarzschild metric (12.13)

12.3.1 Is AdS Black Hole Spacetime Learned?

First, to see if such a learning system works really well, let us check by numerical experiments whether the known metrics are reproduced. Among asymptotically AdS **black holes**, the best known as a solution of the pure Einstein equation is the AdS Schwarzschild solution. For $d = 3$, it is a simple function (using the unit of $R = 1$),

$$h(\eta) = 3\coth(3\eta)\,. \tag{12.13}$$

Now, will this be reproduced by learning?

We discretize the η direction of the bulk into 10 layers and select cutoffs as $\eta_{\text{ini}} = 1$ and $\eta_{\text{fin}} = 0.1$. The mass and potential terms in the equation of motion are chosen as $m^2 = -1$ and $V[\phi] = \frac{1}{4}\phi^4$ for simplicity. First, we produce a set of "correct data" using the discretized AdS Schwarzschild metric (12.13). From the randomly generated $(\phi(\eta_{\text{ini}}), \pi(\eta_{\text{ini}}))$, we use the neural network with the weight of (12.13), and name $(\phi(\eta_{\text{ini}}), \pi(\eta_{\text{ini}}))$ that satisfies $|\pi| < 0.1$ at the final layer as the "positive data." Similarly, those that do not meet the conditions at the final layer are called "negative data." We collect 1000 such data to make a training data set with a positive/negative label. See Fig. 12.3.

Next, using this training data, we start from a random initial set of weight $h(\eta^{(n)})$ and train the neural network.[9] The result is shown in Fig. 12.4. The plots (a-1)(a-2) are the state before the training, and (b-1)(b-2) are the state after the training. The figures on the left show how well the data is judged to be correct. The figures on the right are the plots of the metric function $h(\eta)$. As the training progresses, it can be seen that positive data and negative data can be correctly discriminated. At the

[9] PyTorch as a Python library for learning was used. The figure shows the result of 100 epochs learning with a batch size of 10.

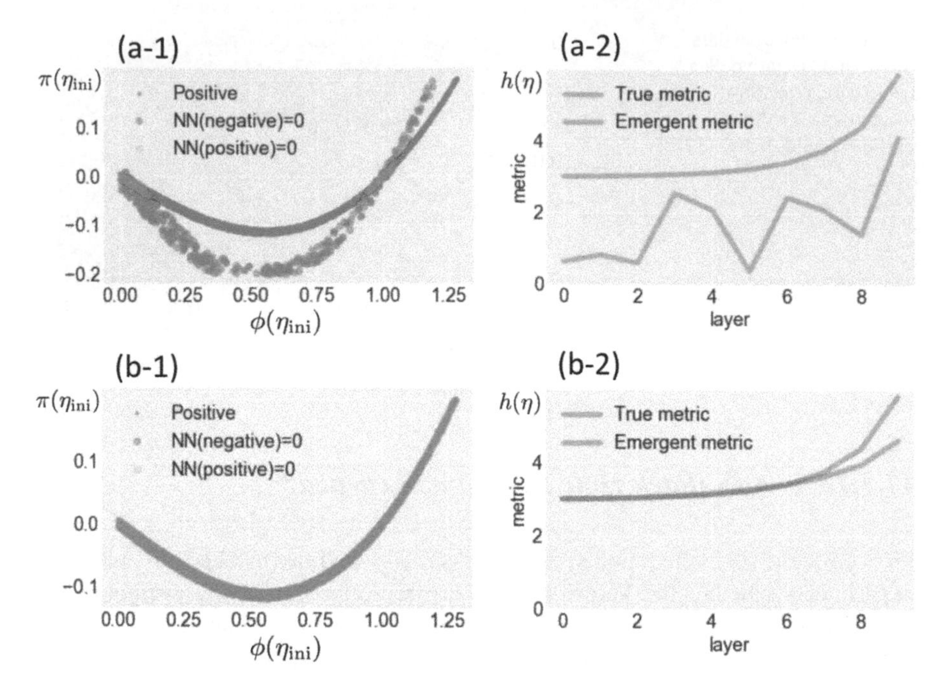

Fig. 12.4 Results of the AdS Schwarzschild spacetime reproduction experiment. In the upper row, (a-1) and (a-2) show the discrimination data and metric before the training, and in the lower row, (b-1) and (b-2) show the data after the training. In (a-1), blue dots are the positive data, and orange dots are the data judged to be positive by the randomly generated initial metric (orange zigzag line in figure (a-2)). These are, of course, different from each other. On the other hand, they match in (b-1) after the training. According to (b-2), the metric after the training (described as the "emergent metric") reproduces the AdS Schwarzschild spacetime (described as the "true metric") except for a small region where η is close to 0

same time, you can see how the resulting metric reproduces the AdS Schwarzschild solution (12.13).

It is necessary to introduce appropriate regularization during this numerical experiment. If you do not introduce regularization, the resulting configuration of $h(\eta)$ is often jagged. Even with the jagged configuration of the metric, the network can distinguish the positive and the negative data. This is because, in general, the weights of a neural network after the training are not unique, and various local minima of the error function are obtained as training results. Among the various $h(\eta)$ obtained, to pick up a smooth metric that is meaningful as a spacetime, we add the following regularization term:

$$L_{\rm reg}^{(1)} \equiv c_{\rm reg} \sum_{n=1}^{N-1} (\eta^{(n)})^4 \left(h(\eta^{(n+1)}) - h(\eta^{(n)}) \right)^2 . \tag{12.14}$$

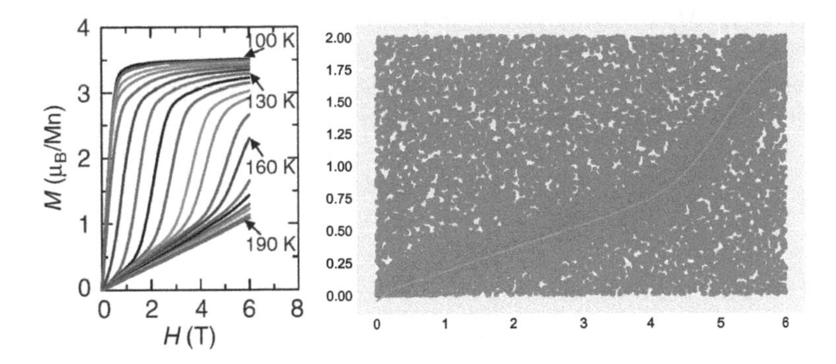

Fig. 12.5 Left: For a manganese oxide $Sm_{0.6}Sr_{0.4}MnO_3$, magnetization $M[\mu_B/M_n]$ against external magnetic field H [Tesla] plotted at various temperatures. Excerpt from the paper [140]. Right: positive data and negative data generated by adding thickness to the 155K data

By adding this term, the values of h in the adjacent η become closer, and a smooth metric can be obtained. This term works to reduce the error function $\int d\eta\,(h'(\eta)\eta^2)^2$ in the continuous limit. Numerical experiments show that if the coefficient of the added regularization is $c_{\text{reg}} = 10^{-3}$, smooth metrics can be extracted without increasing the final loss value.

This numerical experiment shows that the AdS Schwarzschild spacetime can be reproduced by solving the inverse problem from the one-point function of the operator on the boundary side. Deep learning has been found to be effective in solving the inverse problem.

12.3.2 Emergent Spacetime from Material Data

Next, let us take a look at what spacetime emerges when the data of a real material is adopted as the data of the one-point function. What is often measured as a one-point function in matter is the external magnetic field response. In particular, what is called **strongly correlated matter** has a strong correlation between electrons and spins, and is compatible with AdS/CFT that considers the strong coupling limit on the boundary side. Therefore, we will use the data of the external magnetic field response of a manganese oxide $Sm_{0.6}Sr_{0.4}MnO_3$ [140]. In Fig. 12.5, we pick up the data at the temperature of 155K on the left, and divide the data into positive data and negative data (Fig. 12.5 right).[10]

Next, we consider the input part of the neural network. Since the boundary value is related to a one-point function of quantum field theory, such as (12.9) and (12.10),

---

[10]As the experimental data has no error bar, we added the thickness of the data by hand. The technical reason for it is that the training does not progress when the data width is too small.

the data of (H, M) of matter can be converted to (ϕ, π). However, in the present case, the fact that the AdS radius R of the matter is unknown, and the normalization of the operators that must usually be determined from the two-point function of M is unknown, we shall reluctantly put it as follows:

$$\begin{aligned} \phi(\eta_{\text{ini}}) &= \alpha H + \beta M\,, \\ \pi(\eta_{\text{ini}}) &= -\Delta_{-}\alpha H - \Delta_{+}\beta M\,. \end{aligned} \tag{12.15}$$

Here α and β are unknown normalization constants, and

$$\Delta_{\pm} = (d/2)\left(1 \pm \sqrt{1 + 4m^2/h(\infty)^2}\right) \tag{12.16}$$

determines the conformal dimension of the operator ($d/h(\infty)$ gives the AdS radius). On the numerical code, everything is measured in the unit $R_{\text{unit}} = 1$. We consider this (12.15) as the 0th layer and add it to the original neural network made of $h(\eta)$.

The input data and output data are handled in the same way as in the previous AdS Schwarzschild spacetime reproduction experiment, and the weight of the neural network is trained.[11] In the 10-layer neural network, the parameters to be trained in this case are the bulk mass m, λ in the interaction potential $V = (\lambda/4)\phi^4$, the normalization constant α and β, and the metric $h(\eta)$.

Figure 12.6 shows the training result. The left and middle figures show the fitting of the positive data after the training and the emergent metric $h(\eta)$, respectively. You can see that a smooth emergent metric is obtained.[12] The figure on the right shows statistical data from 13 trials. We can see how it converges to a certain function form. At the same time, the value of the bulk mass and the strength of the interaction are learned. The result is $m^2R^2 = 5.6 \pm 2.5$ and $\lambda/R = 0.61 \pm 0.22$.

In this way, by using a deep neural network, the inverse problem for AdS/CFT can be solved. From the data of a given material, interesting questions await, such as what properties the emergent spacetime has and how generally it can be used for AdS/CFT correspondence.

---

[11] In this numerical experiment, we perform the training with two regularization terms: $L_{\text{reg}} = L^{(1)}_{\text{reg}} + L^{(2)}_{\text{reg}}$. The first one $L^{(1)}_{\text{reg}}$ is the same as the previous (12.14). For the second term, we introduced the regularization

$$L^{(2)}_{\text{reg}} \equiv c^{(2)}_{\text{reg}}(h(\eta^{(N)}) - 1/\eta^{(N)})^2\,. \tag{12.17}$$

This makes $h(\eta^{(N)})$ behave like the horizon condition $h(\eta) \sim 1/\eta$ near the black hole horizon. Examining the size of the regularization term so as not to hinder the training, we adopted the value $c^{(2)}_{\text{reg}} = 10^{-4}$ as its coefficient.

[12] We stopped the training when the total error function went below 0.02. This is based on the judgment that the training result is sufficiently close to the training data.

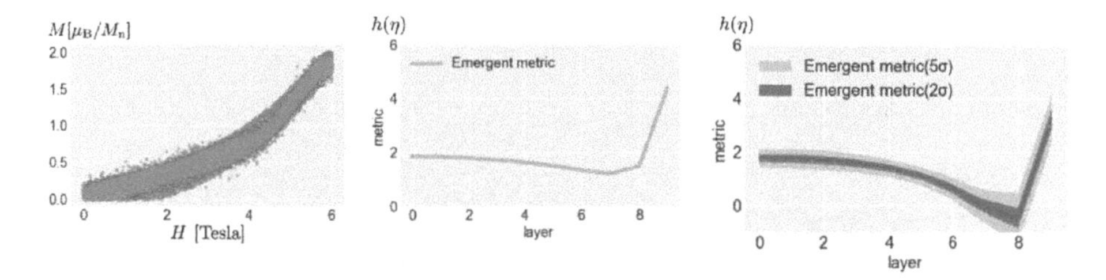

Fig. 12.6 The spacetime metric which emerged from the magnetic field response of manganese oxide $Sm_{0.6}Sr_{0.4}MnO_3$. Middle: Emergent metric function $h(\eta)$. Left: Comparison of magnetic field response (green + orange) calculated with the emergent metric, and actual magnetic field response data (green + blue). Right: Emergent metric, its mean and variance of statistical data from 13 trials

Table 12.1 Left: Chiral condensate [141] as a function of quark mass (in units of lattice constant a). Right: The same data translated into physical units (using $1/a = 0.829(19)$ [GeV]). The error is about 8%

m_q	$\langle\bar{\psi}\psi\rangle$
0.0008125	0.0111(2)
0.0016250	0.0202(4)
0.0032500	0.0375(5)
0.0065000	0.0666(8)
0.0130000	0.1186(5)
0.0260000	0.1807(4)

m_q[GeV]	$\langle\bar{\psi}\psi\rangle$ [(GeV)^3]
0.00067	0.0063
0.0013	0.012
0.0027	0.021
0.0054	0.038
0.011	0.068
0.022	0.10

12.4 Spacetime Emerging from QCD

Then, finally, let us obtain the spacetime emerging from the data of QCD [130], by solving the inverse problem, regarding the asymptotically AdS spacetime as a neural network, as before.

The most important one-point function in QCD is the chiral condensate $\langle\bar{q}q\rangle$, where $q(x)$ is the quark field.[13] The chiral condensate is a vacuum expectation value of the operator $\bar{q}q$, while $\bar{q}q$ is included in the QCD Lagrangian as the quark mass term $m_q\bar{q}q$. So, the source of $\bar{q}q$ is the quark mass m_q. Fortunately, there exists data of the chiral condensate $\langle\bar{q}q\rangle$ measured by lattice QCD while changing m_q. Let us use that as our training data. Table 12.1 gives $\langle O\rangle_g = \langle\bar{q}q\rangle_{m_q}$ measured in physical units [141]. The temperature is chosen to be 207 [MeV].

Next, take the area of $0 < m_q < 0.022$ and $0 < \langle\bar{q}q\rangle < 0.11$, and generate random points in that region. Regard points located close to the curve (which is a cubic function in m_q fitting the data of Table 12.1) within the distance 0.004 as positive data, and points that do not enter the curve neighborhood as negative data (Fig. 12.7). 10000 points are randomly generated and used as our training data.

[13] The chiral condensate corresponds to the magnetization in the Ising model. See Chap. 8.

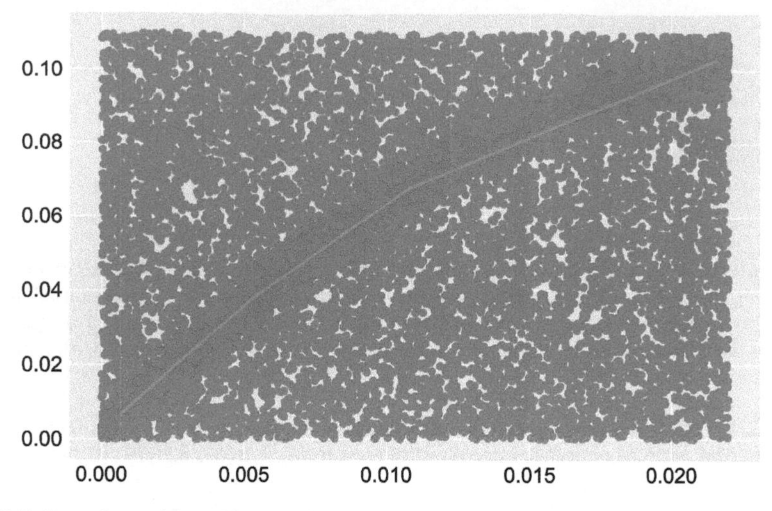

Fig. 12.7 Data for training. The x axis is the quark mass [GeV] and the y axis is the chiral condensate [GeV^3], where the blue dots indicate positive data and the orange dots indicate negative data

To map the data to asymptotic information of the scalar field ϕ in the asymptotic AdS spacetime, we again use the AdS/CFT dictionary [139]. Scalar field theory in the bulk is the ϕ^4 theory, and the potential is

$$V[\phi] = \frac{\lambda}{4}\phi^4. \tag{12.18}$$

This coupling constant λ is also subject to learning (learned values must be positive). Since the dimension of the chiral condensate operator $\overline{q}q$ is 3, the formula

$$\Delta_{\mathcal{O}} \equiv (d/2) + \sqrt{d^2/4 + m^2 R^2} \tag{12.19}$$

gives the mass of the scalar field as $m^2 = -3/R^2$. Here R is the AdS radius, which is also the subject of learning. At the asymptotic boundary of AdS, $h(\eta) \approx 4/R$, so the equation of motion of the bulk scalar field there is

$$\partial_\eta^2 \phi + \frac{4}{R}\partial_\eta \phi - \frac{3}{R^2}\phi - \lambda\phi^3 = 0\,. \tag{12.20}$$

The solution is as follows:

$$R^{3/2}\,\phi \approx \alpha e^{-\eta/R} + \beta e^{-3\eta/R} - \frac{\lambda\alpha^3}{2R^2}\eta\, e^{-3\eta/R}\,. \tag{12.21}$$

Here, the first term is a non-normalizable mode, the second term is a normalizable mode, and the third term is a term that comes from the interaction term. The coefficients are related to the quark mass and the chiral condensate as follows:

$$\alpha = \frac{\sqrt{N_c}}{2\pi} m_q R\,, \quad \beta = \frac{\pi}{\sqrt{N_c}} \langle \bar{q} q \rangle R^3\,. \tag{12.22}$$

Since the data is that of QCD, we substitute $N_c = 3$ hereafter. All numerical calculations will be measured in units of R and handle dimensionless quantities. In summary, the QCD data is mapped to the value of the scalar field at the asymptotic boundary,

$$\phi(\eta_{\text{ini}}) = \alpha e^{-\eta_{\text{ini}}} + \beta e^{-3\eta_{\text{ini}}} - \frac{\lambda\alpha^3}{2} \eta_{\text{ini}}\, e^{-3\eta_{\text{ini}}}. \tag{12.23}$$

$$\pi(\eta_{\text{ini}}) = -\alpha e^{-\eta_{\text{ini}}} - \left(3\beta + \frac{\lambda\alpha^3}{2}\right) e^{-3\eta_{\text{ini}}} + \frac{3\lambda\alpha^3}{2} \eta_{\text{ini}} e^{-3\eta_{\text{ini}}}\,. \tag{12.24}$$

The unknown functions determined in the training are λ, R, $h(\eta)$.

We will use a 15-layer neural network and discretize the η direction accordingly. If the ends of the spacetime are chosen as $\eta_{\text{ini}} = 1$ and $\eta_{\text{fin}} = 0.1$ in units of $R = 1$, then $\eta_{\text{fin}} \leq \eta \leq \eta_{\text{ini}}$ will be equally divided into 15 parts. With this setup, we performed machine learning.[14] The condition for discriminating positive and negative data is the same event horizon condition as before. We implemented machine learning using PyTorch. As an initial condition, we took h as randomly generated ones around $h = 4$ with a fluctuation of size 3, $\lambda = 0.2$ and $R = 0.8$ [GeV^{-1}]. The batch size was 10, and the training was completed at 1500 epochs. The result of the training (8 trials) with the error less than 0.008 is shown in Table 12.2, and the plot is shown in Fig. 12.8. The learned coupling constant λ and

[14] We use the regularization $L_{\text{reg}} = L_{\text{reg}}^{(\text{smooth})} + L_{\text{reg}}^{(\text{bdry})}$. The first term is

$$L_{\text{reg}}^{(\text{smooth})} \equiv c_{\text{reg}} \sum_{n=1}^{N-1} (\eta^{(n)})^4 \left(h(\eta^{(n+1)}) - h(\eta^{(n)})\right)^2\,, \tag{12.25}$$

which makes $h(\eta)$ a smooth function ($c_{\text{reg}} = 0.01$). The factor η^4 does not prohibit the function form $h(\eta) \propto 1/\eta$ expected near the event horizon. The second term is

$$L_{\text{reg}}^{(\text{bdry})} \equiv c_{\text{reg}} \left(d - h(\eta^{(1)})\right)^2\,, \tag{12.26}$$

which is to make sure that the asymptotic spacetime is $(d+1)$-dimensional AdS spacetime. The coefficient was chosen as $c_{\text{reg}} = 0.01$.

Table 12.2 The spacetime metric $h(\eta^{(n)})$ ($n-1 = 0, 1, \cdots, 14$ is the layer number) which emerged after the training

$n-1$	h
0	3.0818 ± 0.0081
1	2.296 ± 0.016
2	1.464 ± 0.025
3	0.627 ± 0.035
4	-0.141 ± 0.045
5	-0.727 ± 0.049
6	-0.974 ± 0.043
7	-0.687 ± 0.032

n-1	h
8	0.374 ± 0.087
9	2.50 ± 0.19
10	6.03 ± 0.30
11	11.46 ± 0.35
12	19.47 ± 0.27
13	31.07 ± 0.17
14	46.70 ± 0.52

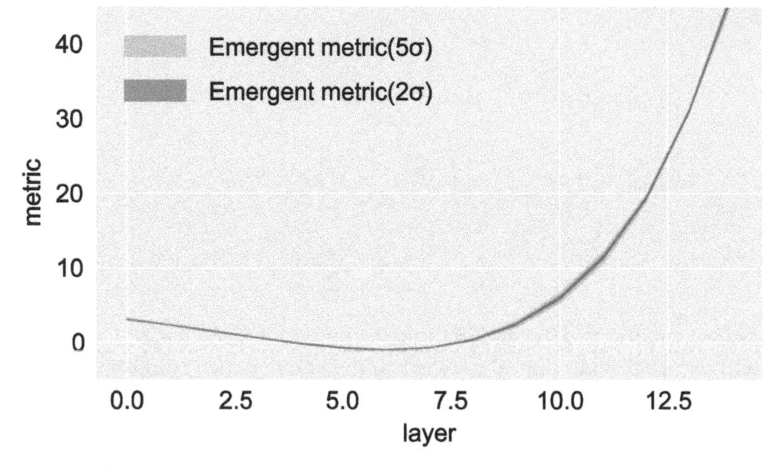

Fig. 12.8 A plot of the emergent spacetime metric (Table 12.2). The y axis is the metric $h(\eta)$, and the x axis is the emergent spatial coordinate η, that is, the layer number $n-1 = 0, \cdots, 14$

AdS radius R are

$$\lambda = 0.01243 \pm 0.00060\,, \tag{12.27}$$

$$R = 3.460 \pm 0.021[\text{GeV}^{-1}]\,. \tag{12.28}$$

This means that, with the standard conversion 5.0677 $[\text{GeV}^{-1}] = 1$ [fm], we obtain $R = 0.6828 \pm 0.0041$ [fm].

Let us look at the metric that emerged from the QCD data. There are three interesting points:

- All eight cases have given the same emergent metric. Therefore, the obtained metrics can be said to be universal.
- The resulting $h(\eta^{(n)})$ diverges as η approaches the event horizon. This is the same behavior as the black hole metric. It is considered to be a black hole which automatically emerged from the data of finite temperature lattice QCD.

- In the middle region, $h(\eta)$ is negative. This is a behavior not found in any solution of the ordinary vacuum Einstein equation.

In this way, an asymptotically AdS emergent spacetime metric that consistently reproduces the chiral condensate data of QCD was obtained. A method of solving the inverse problem and obtaining a gravity model was made possible by machine learning.

Then, does the metric gained from this learning make any interesting new predictions? The strange aspect about the resulting metric is that h is negative, as mentioned in the third point above. This means that the black hole spacetime actually shows a confinement behavior. The confinement spacetime is a spacetime in which the value of the metric increases at a certain spacetime location, so when going from the asymptotic AdS region to the depth of the spacetime, one reaches the bottom and cannot proceed further. Since h is with a differentiation by η, the reversal of its sign indicates the behavior of such a metric. Therefore, although our spacetime has a black hole horizon (deconfinement), at the same time the metric shows the confinement.

In fact, with this metric, using the AdS/CFT dictionary we can calculate the vacuum expectation value of an operator called the Wilson loop. Then we can see both the properties of Debye screening coming from the event horizon and the linear potential coming from confining spacetime. They qualitatively match well the predictions of lattice QCD. Surprisingly, the spacetime that emerged from QCD data using deep learning had both properties. In this way, when the inverse problem is specifically solved using deep learning, it has become possible to construct a model that gives a concept that goes beyond the conventional holographic model.

It is very interesting that the network itself can be interpreted as a smooth spacetime, beyond the construction of such a practical model. As mentioned in the column of the Ising model, Hopfield models regard the neural network itself as a spin system, and the weight of the network corresponded to the strength of interaction between spins. In the model introduced in this chapter, the network weights directly include the metric. Therefore, the distance between the nodes of the network is defined by the weights, and the whole network is interpreted as a space. Such an idea is especially reminiscent of quantum gravity theory based on dynamical triangulation. How far can the idea of a network becoming spacetime be generalized?

Holographic models based on deep learning are classical gravity theories, but in the future, we may be able to understand how quantum effects and essential effects of gravity appear, and how this modeling is related to other quantum gravity concepts.

Column: Black Holes and Information

In Chap. 1 we explained that Maxwell's demon made an entropy decrease by $\log 2$. Here, as a final column of this book, we shall talk about the introduction of **black hole entropy** by J. Bekenstein and S. Hawking. Prior to their proposal, it was pointed out that there was a similarity of the law of increasing entropy to a black hole growing endlessly by swallowing gas around it. Bekenstein assumed that the entropy S is given as

$$S = f(A)\,, \tag{12.29}$$

where the area of the black hole horizon is A, and attempted to determine the function f. The key in his idea is "What is the minimum amount of change in S?" He assumed the answer as follows:

1. It should correspond to the case where a particle with its Compton length equal to its radial size falls into a black hole.
2. The entropy change should be $\log 2$ since the particle is either kept retained or destroyed.

First, the minimum area change from Assumption 1 is semi-classically calculated as

$$\delta A = 2\hbar. \tag{12.30}$$

Combining this with Assumption 2 results in

$$\log 2 = \delta S = \delta A \cdot f'(A) = 2\hbar f'(A)\,. \tag{12.31}$$

In this manner Bekenstein derived

$$S = f(A) = \frac{\log 2}{2}\frac{A}{\hbar}\,. \tag{12.32}$$

This is Bekenstein's original claim. He mentioned in his paper [6] that "Nevertheless, it should be clear that if the coefficient is not exactly $\frac{\log 2}{2}$, then it must be very close to this, probably within a factor of two." The precise formula obtained by Hawking later, in the paper [7] about black hole evaporation with quantized fields, is

$$S = \frac{1}{4}\frac{A}{\hbar}\,. \tag{12.33}$$

Each of these coefficients is numerically

$$\frac{\log 2}{2} = 0.34\ldots, \qquad (12.34)$$

$$\frac{1}{4} = 0.25\,. \qquad (12.35)$$

They are close to each other. The value log 2 is the approximate information that the particle which fell into the black hole had, and it was converted to the area of the black hole.

In the topic of Maxwell's demon mentioned in Chap. 1 and in this topic of the black hole, the fundamental connection between information and physics is seen. In fact, in recent developments in quantum gravity theory, the idea that spacetime itself is formed by quantum entanglement has become mainstream. Also, while research on the holographic principle is built on the assertion that black holes are the most chaotic, as described in the column in Chap. 4, computation can be formulated at the edge of chaos.

Deep learning has a potential to be the foundation of "intelligence" handling information. In this book, we have seen physics taking an active role there. So far, deep learning and physics seem mutually related in various ways. We are really looking forward to the progress and the evolution of information, computability, machine learning, and physics, in the future.

Chapter 13
Epilogue

Abstract In this book, we have looked at the relationship between deep learning and physics in detail. We authors (Akinori Tanaka, Akio Tomiya, and Koji Hashimoto) are also co-authors of research papers in this interdisciplinary field. Collaborative research is something that is only fun if people who share the same ambition but have different backgrounds gather. We guess that readers have started reading this book with various motivations, but in fact, all three of us have different ideas about the "deep learning and physics." Here, the three authors express their feelings, rather than adopting the concluding style of ordinary textbooks, so that they can be a source of your debate in this new field in the future.

So far, we have looked at the relationship between deep learning and physics in detail. The authors (Akinori Tanaka, Akio Tomiya, and Koji Hashimoto) are also co-authors of research papers in this interdisciplinary field. Collaborative research is something that is only fun if people who share the same ambition but have different backgrounds gather. We guess the readers have started reading this book with various motivations, but in fact, all three of us have different ideas about the "deep learning and physics." Here, the three authors express their feelings, rather than adopting the concluding style of ordinary textbooks, so that they can be a source of your debate in this new field in the future.

13.1 Neural Network, Physics and Technological Innovation (Akio Tomiya)

From 2003 to 2005, when I was a junior high school student, the world was in the midst of the second winter of AI. As a computer boy, I was making computer games, programs—a tic-tac-toe program beating humans, and chatbots.[1] Speaking

[1]Chatbots are programs which react automatically to specific keywords.

A. Tanaka et al., *Deep Learning and Physics*, Mathematical Physics Studies,
https://doi.org/10.1007/978-981-33-6108-9_13

of artificial intelligence at that time, there were expert systems and IBM's Deep Blue.[2] While doing various searches using the Google webpage (which was already there at that time), I learned that neural networks were terribly interesting, but gave up because I couldn't understand the backpropagation method. On the other hand, after entering high school, I began to find physics interesting, and then in 2015 I got a PhD in particle physics and became a postdoc researcher. Then big news came in 2016 that Google's artificial intelligence had defeated humans at Go.[3] In response to the news, I began study and research while exchanging messages with one of the authors of this book, Akinori Tanaka. When I studied machine learning for the first time, I realized that mathematical tools were similar to physics, so I could easily understand ideas and applied them. As I studied further, I understood that some ideas were imported from physics.[4] The research went well and was formed into several papers. In this book, physics researchers explain deep learning and related fields from the basics to the application from that point of view, and in fact, this is the path I followed.

My specialty is lattice QCD, and it was no surprise to discover that a person in the field is one of the developers of Jupyter Notebook.[5] This is because if you are doing physics using a computer, you have to visualize a large amount of data. The WWW, the core technology of today's Internet, was developed at the European Nuclear Research Organization (CERN).[6] Also, the preprint service called arXiv[7] was created by a particle physicist.[8] At the meta level, physics supports the next generation of technology. I am glad if the readers who have studied physics and deep learning through this book will theoretically solve the mystery of generalization of deep learning and clarify the relationship between physics and machine learning. At the same time, I would be happy as well if readers will be responsible for the next generation of technological innovation that gave us the WWW and Jupyter Notebook.

---

[2]Deep Blue was a supercomputer that defeated the world chess champion Garli Kasparov in 1997.

[3]Deep Mind's Alpha Go defeated Lee Sedol.

[4]This is about the Boltzmann machines that triggered the huge progress in deep learning.

[5]This person is Fernando Pérez from University of Colorado at Boulder. He obtained a PhD in Lattice QCD, and was a student of Anna Hasenfratz who is famous in the field.

[6]The actual development was done by Tim Berners-Lee, an information scientist who was there, but I don't think it would have been done without the accelerator, which is a machine that generates a lot of data by itself, or without a strong motivation to understand nature. It's also not hard to see how the hacker spirit of the physicist community (in the original sense) could make this system free to use.

[7]The arXiv, https://arxiv.org is a web archive service for researchers to upload research preprints on physics and mathematics.

[8]Paul Ginsparg, a prominent physicist whose name is known in the important Gisparg–Wilson relation in lattice QCD, first started the server for the arXiv. The preceding service was founded by a physicist, Joanne Cohn. (I thank Shinichi Nojiri, Professor of Nagoya University, for providing this information.)

13.2 Why Does Intelligence Exist? (Akinori Tanaka)

Rachel Carson, who is one of the most famous biologists and wrote a book *Silent Spring* which describes the harmful effects of chemicals on ecosystems, has a short essay in her other book, *The Sense of Wonder*. In that essay, the sense of wonder is "the mind that feels the nature as it is and as being wonderful." When learning / studying physics and mathematics, people often get asked "why such a useless subject?" or are labeled as "freakish." There were such times for me, and I was frustrated when I couldn't verbalize why I got attracted to these fields. But a possible answer may be, to feel the sense of wonder.

Why do we have such a "sense of wonder" in the first place? If the brain's actions could all be explained by chemical reactions, there is no possibility that "consciousness" would appear there. Everything would simply be a physical phenomenon caused by chemical reactions. In that sense, consciousness may be an illusion. Even if consciousness is an illusion, at least humans have "intelligence," and there is no doubt that they have ability of logical information processing.

Because there are intelligent creatures in this world, I feel that there may be some relationship between the existence of intelligence and the laws of physics. I'm interested in the great (and naive) mystery of why the universe started, why it's expanding, why it's four-dimensional, etc., but just as much, I think that the following is also an interesting question: why does intelligence exist? Once you know the answer, you may even be able to create intelligence. Although the development in deep learning in recent years has shown its potential, as is often said, there is currently no theoretical explanation why deep learning works. Revealing that, you may be able to approach the mystery of intelligence.

13.3 Why do Physical Laws Exist? (Koji Hashimoto)

"Philosophy is useless when doing physics." When I was a freshman in graduate school, I walked down the hall with Thomas Kuhn's *The Structure of the Scientific Revolution*, and I still remember that my supervisor told me those words. Is it really useless?

"The general coordinate transformation invariance, which is the principle of general relativity, is natural, because it is a way for humans to recognize the outside world." When I was a freshman at university, I was told this by a particle theorist, Sho Tanaka. Is it true?

As discussed in Chap. 7 of this book, the revolutionary discovery of concepts in physics is an inverse problem. What performs learning (which is an inverse problem) is neural networks, which are the means by which humans recognize, think, and remember the outside world. And, as we have seen repeatedly in this book, artificial neural networks use the concept of physics when implemented as deep learning and on machines. Is there something hidden in these circling connections that can be called a “theory of theories”?

In learning the amazing developments in machine learning in recent years, I have been repeatedly aware of its relationship to basic and innovative concepts in physics, the frequency and breadth of which was surprising. Is this feeling a tautology simply due to my working memory being too small?

Reduction theory of elementary particles has a history that it has evolved at the same time as new concepts have been created. I feel that it is impossible for machine learning to be completely unrelated to the notion that the origin of quantum gravity spacetime itself, which is the subject of superstring theory, is ultimately elucidated by human intelligence. I look forward to meeting the readers and discussing physics.

Bibliography

1. Utiyama, R.: Theory of Relativity (in Japanese). Iwanami Shoten (1987)
2. Weinberg, S.: Scientist: four golden lessons. Nature **426**(6965), 389 (2003)
3. Toyoda, T.: Physics of Information (in Japanese). Butsuri-no-Taneakashi. Kodansha (1997)
4. Amari, S.-I.: Information Theory (in Japanese). Chikuma Shobo (2011)
5. Shannon, C.E.: A mathematical theory of communication. Bell Syst. Tech. J. **27**(3), 379–423 (1948)
6. Bekenstein, J.D.: Black holes and entropy. Phys. Rev. D **7**(8), 2333 (1973)
7. Hawking, S.W.: Particle creation by black holes. Commun. Math. Phys. **43**(3), 199–220 (1975)
8. Sagawa, T., Ueda, M.: Information thermodynamics: Maxwell's demon in nonequilibrium dynamics. In: Nonequilibrium Statistical Physics of Small Systems: Fluctuation Relations and Beyond, Wiley Online Library, pp. 181–211 (2013)
9. Wheeler, J.A.: Information, physics, quantum: The search for links. In: Proceedings of the 3rd International Symposium on Foundations of Quantum Mechanics, Tokyo, 1989, pp. 354–368, Physical Society of Japan (1990)
10. Nielsen, M.A., Chuang, I.L.: Quantum Computation and Quantum Information. Cambridge University Press (2010)
11. Witten, E.: A mini-introduction to information theory. arXiv preprint arXiv:1805.11965 (2018)
12. Sanov, I.N.: On the probability of large deviations of random magnitudes. Matematicheskii Sbornik **84**(1), 11–44 (1957)
13. Sanov, I.N.: On the Probability of Large Deviations of Random Variables. United States Air Force, Office of Scientific Research (1958)
14. Rosenberg, C.: The lenna story. www.lenna.org
15. Poundstone, W.: The Recursive Universe. Contemporary Books (1984)
16. Samuel, A.L.: Some studies in machine learning using the game of checkers. II – recent progress. In: Computer Games I, pp. 366–400. Springer (1988)
17. Feynman, R.P., Leighton, R.B., Sands, M.: The Feynman Lectures on Physics: Mainly Electromagnetism and Matter, vol. 2. Addison-Wesley, Reading. reprinted (1977)
18. LeCun, Y.: The mnist database of handwritten digits. http://yann.lecun.com/exdb/mnist/ (1998)
19. Krizhevsky, A., Hinton, G.: Learning multiple layers of features from tiny images. Technical report, Citeseer (2009)
20. Bishop, C.M.: Pattern Recognition and Machine Learning. Springer (2006)

A. Tanaka et al., *Deep Learning and Physics*, Mathematical Physics Studies,
https://doi.org/10.1007/978-981-33-6108-9

21. Mohri, M., Rostamizadeh, A., Talwalkar, A.: Foundations of Machine Learning. The MIT Press (2018)
22. Akaike, H.: Information theory and an extension of the maximum likelihood principle. In: Selected Papers of Hirotugu Akaike, pp. 199–213. Springer (1998)
23. Borsanyi, S., et al.: Ab initio calculation of the neutron-proton mass difference. Science **347**, 1452–1455 (2015)
24. He, K., Zhang, X., Ren, S., Sun, J.: Deep residual learning for image recognition. In: Proceedings of the IEEE Conference on Computer Vision and Pattern Recognition, pp. 770–778 (2016)
25. Shalev-Shwartz, S., Ben-David, S.: Understanding Machine Learning: From Theory to Algorithms. Cambridge University Press (2014)
26. Deng, J., Dong, W., Socher, R., Li, L., Li, K., Fei-Fei, L.: Imagenet: a large-scale hierarchical image database. In: 2009 IEEE Conference on Computer Vision and Pattern Recognition, pp. 248–255 (2009)
27. Kawaguchi, K., Kaelbling, L.P., Bengio, Y.: Generalization in deep learning. arXiv preprint arXiv:1710.05468 (2017)
28. Goodfellow, I., Bengio, Y., Courville, A.: Deep Learning. MIT Press (2016)
29. Nair, V., Hinton, G.E.: Rectified linear units improve restricted boltzmann machines. In: Proceedings of the 27th International Conference on Machine Learning (ICML-10), pp. 807–814 (2010)
30. Teh, Y.W., Hinton, G.E.: Rate-coded restricted boltzmann machines for face recognition. In: Advances in Neural Information Processing Systems, pp. 908–914 (2001)
31. Rumelhart, D.E., Hinton, G.E., Williams, R.J.: Learning internal representations by error propagation. Technical report, California Univ San Diego La Jolla Inst for Cognitive Science (1985)
32. Cybenko, G.: Approximation by superpositions of a sigmoidal function. Math. Control Signals Syst. **2**(4), 303–314 (1989)
33. Nielsen, M.: A visual proof that neural nets can compute any function. http://neuralnetworksanddeeplearning.com/chap4.html.
34. Lee, H., Ge, R., Ma, T., Risteski, A., Arora, S.: On the ability of neural nets to express distributions. arXiv preprint arXiv:1702.07028 (2017)
35. Sonoda, S., Murata, N.: Neural network with unbounded activation functions is universal approximator. Appl. Comput. Harmon Anal. **43**(2), 233–268 (2017)
36. LeCun, Y., Bottou, L., Bengio, Y., Haffner, P., et al.: Gradient-based learning applied to document recognition. Proc. IEEE **86**(11), 2278–2324 (1998)
37. Hubel, D.H., Wiesel, T.N.: Receptive fields, binocular interaction and functional architecture in the cat's visual cortex. J. Physiol. **160**(1), 106–154 (1962)
38. Fukushima, K., Miyake, S.: Neocognitron: a new algorithm for pattern recognition tolerant of deformations and shifts in position. Pattern Recognit. **15**(6), 455–469 (1982)
39. Krizhevsky, A., Sutskever, I., Hinton, G.E.: Imagenet classification with deep convolutional neural networks. In: Advances in Neural Information Processing Systems, pp. 1097–1105 (2012)
40. Sabour, S., Frosst, N., Hinton, G.E.: Dynamic routing between capsules. In: Advances in Neural Information Processing Systems, pp. 3856–3866 (2017)
41. Dumoulin, V., Visin, F.: A guide to convolution arithmetic for deep learning. arXiv preprint arXiv:1603.07285 (2016)
42. Radford, A., Metz, L., Chintala, S.: Unsupervised representation learning with deep convolutional generative adversarial networks. arXiv preprint arXiv:1511.06434 (2015)
43. Siegelmann, H.T., Sontag, E.D.: On the computational power of neural nets. J. Comput. Syst. Sci. **50**(1), 132–150 (1995)
44. Hochreiter, S.: Untersuchungen zu dynamischen neuronalen Netzen. Diploma thesis, Institut für Informatik (1991)
45. Hochreiter, S., Bengio, Y., Frasconi, P., Schmidhuber, J., et al.: Gradient flow in recurrent nets: the difficulty of learning long-term dependencies. In: A Field Guide to Dynamical Recurrent

Neural Networks. IEEE Press (2001)
46. Hochreiter, S., Schmidhuber, J.:. Long short-term memory. Neural Comput. **9**(8), 1735–1780 (1997)
47. Olah, C.: Understanding LSTM networks (2015). http://colah.github.io/posts/2015-08-Understanding-LSTMs/
48. Bahdanau, D., Cho, K., Bengio, Y.: Neural machine translation by jointly learning to align and translate. arXiv preprint arXiv:1409.0473 (2014)
49. Vaswani, A., Shazeer, N., Parmar, N., Uszkoreit, J., Jones, L., Gomez, A.N., Kaiser, Ł., Polosukhin, I.: Attention is all you need. In: Advances in Neural Information Processing Systems, pp. 5998–6008 (2017)
50. Xu, K., Ba, J., Kiros, R., Cho, K., Courville, A., Salakhudinov, R., Zemel, R., Bengio, Y.: Show, attend and tell: neural image caption generation with visual attention. In: International Conference on Machine Learning, pp. 2048–2057 (2015)
51. Zhang, H., Goodfellow, I., Metaxas, D., Odena, A.: Self-attention generative adversarial networks. arXiv preprint arXiv:1805.08318 (2018)
52. Olah, C., Carter, S.: Attention and augmented recurrent neural networks. Distill (2016)
53. Graves, A., Wayne, G., Danihelka, I.: Neural turing machines. arXiv preprint arXiv:1410.5401 (2014)
54. Wolfram, S.: Theory and Applications of Cellular Automata: Including Selected Papers 1983–1986. World scientific (1986)
55. Cook, M.: Universality in elementary cellular automata. Complex Syst. **15**(1), 1–40 (2004)
56. Cook, M.: A concrete view of rule 110 computation. arXiv preprint arXiv:0906.3248 (2009)
57. Langton, C.G.: Computation at the edge of chaos: phase transitions and emergent computation. Physica D: Nonlinear Phenomena **42**(1–3), 12–37 (1990)
58. Lyons, L.: Discovering the significance of 5 sigma. arXiv preprint arXiv:1310.1284 (2013)
59. Jona-Lasinio, G.: Renormalization group and probability theory. Phys. Rep. **352**(4–6), 439–458 (2001)
60. Ferrenberg, A.M., Landau, D.P., Wong, Y.J.: Monte carlo simulations: hidden errors from "good" random number generators. Phys. Rev. Lett. **69**(23), 3382 (1992)
61. Matsumoto, M., Nishimura, T.: Mersenne twister: a 623-dimensionally equidistributed uniform pseudo-random number generator. ACM Trans. Model. Comput. Simul. (TOMACS) **8**(1), 3–30 (1998)
62. Savvidy, G.K., Ter-Arutyunyan-Savvidy, N.G.: On the monte carlo simulation of physical systems. J. Comput. Phys. **97**(2), 566–572 (1991)
63. Marsaglia, G., et al.: Xorshift RNGS. J. Stat. Softw. **8**(14), 1–6 (2003)
64. Box, G.E.P.: A note on the generation of random normal deviates. Ann. Math. Stat. **29**, 610–611 (1958)
65. Häggström, O.: Finite Markov Chains and Algorithmic Applications, vol. 52. Cambridge University Press (2002)
66. Tierney, L.: Markov chains for exploring posterior distributions. Ann. Stat. **22**(4), 1701–1728 (1994)
67. Chowdhury, A., De, A.K., De Sarkar, S., Harindranath, A., Maiti, J., Mondal, S., Sarkar, A.: Exploring autocorrelations in two-flavour Wilson Lattice QCD using DD-HMC algorithm. Comput. Phys. Commun. **184**, 1439–1445 (2013)
68. Luscher, M.: Computational strategies in Lattice QCD. In: Modern perspectives in lattice QCD: Quantum Field Theory and High Performance Computing. Proceedings, International School, 93rd Session, Les Houches, France, 3–28 Aug 2009, pp. 331–399 (2010)
69. Madras, N., Sokal, A.D.: The pivot algorithm: a highly efficient monte carlo method for the self-avoiding walk. J. Stat. Phys. **50**(1–2), 109–186 (1988)
70. Metropolis, N., Rosenbluth, A.W., Rosenbluth, M.N., Teller, A.H., Teller, E.: Equation of state calculations by fast computing machines. J. Chem. Phys. **21**(6), 1087–1092 (1953)
71. Creutz, M., Jacobs, L., Rebbi, C.: Monte carlo study of abelian lattice gauge theories. Phys. Rev. D **20**(8), 1915 (1979)

72. Geman, S., Geman, D.: Stochastic relaxation, gibbs distributions, and the bayesian restoration of images. IEEE Trans. Pattern Anal. Mach. Intell. PAMI **6**(6), 721–741 (1984)
73. Hastings, W.K.: Monte Carlo sampling methods using Markov chains and their applications. Oxford University Press (1970)
74. Onsager, L.: Crystal statistics. I. A two-dimensional model with an order-disorder transition. Phys. Rev. **65**(3–4), 117 (1944)
75. Nambu, Y.: A note on the eigenvalue problem in crystal statistics. In: Broken Symmetry: Selected Papers of Y. Nambu, pp. 1–13. World Scientific (1995)
76. Sutton, R.S., Barto, A.G., et al.: Introduction to Reinforcement Learning, vol. 135. MIT Press, Cambridge, MA (1998)
77. Hinton, G.E.: Training products of experts by minimizing contrastive divergence. Neural Comput. **14**(8), 1771–1800 (2002)
78. Bengio, Y., Delalleau, O.: Justifying and generalizing contrastive divergence. Neural Comput. **21**(6), 1601–1621 (2009)
79. Goodfellow, I., Pouget-Abadie, J., Mirza, M., Xu, B., Warde-Farley, D., Ozair, S., Courville, A., Bengio, Y.: Generative adversarial nets. In: Advances in Neural Information Processing Systems, pp. 2672–2680 (2014)
80. Neumann, V.: Zur theorie der gesellschaftsspiele. Mathematische Annalen **100**(1), 295–320 (1928)
81. Zhao, J., Mathieu, M., LeCun, Y.: Energy-based generative adversarial network. arXiv preprint arXiv:1609.03126 (2016)
82. Arjovsky, M., Chintala, S., Bottou, L.: Wasserstein gan. arXiv preprint arXiv:1701.07875 (2017)
83. Villani, C.: Optimal Transport: Old and New, vol. 338. Springer Science & Business Media (2008)
84. Peyré, G., Cuturi, M., et al.: Computational optimal transport. Found. Trends® Mach. Learn. **11**(5–6), 355–607 (2019)
85. Cuturi, M.: Sinkhorn distances: lightspeed computation of optimal transport. In: Advances in Neural Information Processing Systems, pp. 2292–2300 (2013)
86. Gulrajani, I., Ahmed, F., Arjovsky, M., Dumoulin, V., Courville, A.C.: Improved training of wasserstein gans. In: Advances in Neural Information Processing Systems, pp. 5767–5777 (2017)
87. Miyato, T., Kataoka, T., Koyama, M., Yoshida, Y.: Spectral normalization for generative adversarial networks. arXiv preprint arXiv:1802.05957 (2018)
88. Kingma, D.P., Welling, M.: Auto-encoding variational bayes. arXiv preprint arXiv:1312.6114 (2013)
89. Tolstikhin, I., Bousquet, O., Gelly, S., Schoelkopf, B.: Wasserstein auto-encoders. arXiv preprint arXiv:1711.01558 (2017)
90. Dinh, L., Krueger, D., Bengio, Y.: Nice: Non-linear independent components estimation. arXiv preprint arXiv:1410.8516 (2014)
91. Wang, L.: Generative models for physicists (2018). https://wangleiphy.github.io/lectures/piltutorial.pdf. https://wangleiphy.github.io/lectures/PILtutorial.pdf
92. Dodge, S., Karam, L.: A study and comparison of human and deep learning recognition performance under visual distortions. In: 26th International Conference on Computer Communication and Networks (ICCCN), pp. 1–7. IEEE (2017)
93. Barratt, S., Sharma, R.: A note on the inception score. arXiv preprint arXiv:1801.01973 (2018)
94. Salimans, T., Goodfellow, I., Zaremba, W., Cheung, V., Radford, A., Chen, X.: Improved techniques for training gans. In: Advances in Neural Information Processing Systems, pp. 2234–2242 (2016)
95. Heusel, M., Ramsauer, H., Unterthiner, T., Nessler, B., Hochreiter, S.: Gans trained by a two time-scale update rule converge to a local nash equilibrium. In: Advances in Neural Information Processing Systems, pp. 6626–6637 (2017)

96. Szegedy, C., Liu, W., Jia, Y., Sermanet, P., Reed, S., Anguelov, D., Erhan, D., Vanhoucke, V., Rabinovich, A.: Going deeper with convolutions. In: Proceedings of the IEEE Conference on Computer Vision and Pattern Recognition, pp. 1–9 (2015)
97. Coates, A., Ng, A., Lee, H.: An analysis of single-layer networks in unsupervised feature learning. In: Proceedings of the Fourteenth International Conference on Artificial Intelligence and Statistics, pp. 215–223 (2011)
98. Miyato, T., Koyama, M.: cGANs with projection discriminator. arXiv preprint arXiv:1802.05637 (2018)
99. Brock, A., Donahue, J., Simonyan, K.: Large scale GAN training for high fidelity natural image synthesis. arXiv preprint arXiv:1809.11096 (2018)
100. Maeda, S., Aoki, Y., Ishii, S.: Training of Markov chain with detailed balance learning (in Japanese). In: Proceedings of the 19th Meeting of Japan Neural Network Society, pp. 40–41 (2009)
101. Liu, J., Qi, Y., Meng, Z.Y., Fu, L.: Self-learning monte carlo method. Phys. Rev. B **95**(4), 041101 (2017)
102. Otsuki, J., Ohzeki, M., Shinaoka, H., Yoshimi, K.: Sparse modeling approach to analytical continuation of imaginary-time quantum monte carlo data. Phys. Rev. E **95**(6), 061302 (2017)
103. The Event Horizon Telescope Collaboration: First M87 event horizon telescope results. IV. Imaging the central supermassive black hole. Astrophys. J. Lett. **875**, 1, L4 (2019)
104. Montvay, I., Münster, G.: Quantum Fields on a Lattice. Cambridge University Press (1994)
105. Mehta, P., Bukov, M., Wang, C.-H., Day, A.G.R., Richardson, C., Fisher, C.K., Schwab, D.J.: A high-bias, low-variance introduction to machine learning for physicists. Phys. Rep. **810**, 1–124 (2019)
106. Carrasquilla, J., Melko, R.G.: Machine learning phases of matter. Nat. Phys. **13**(5), 431 (2017)
107. Wang, L.: Discovering phase transitions with unsupervised learning. Phys. Rev. B **94**(19), 195105 (2016)
108. Tanaka, A., Tomiya, A.: Detection of phase transition via convolutional neural networks. J. Phys. Soc. Jpn. **86**(6), 063001 (2017)
109. Kashiwa, K., Kikuchi, Y., Tomiya, A.: Phase transition encoded in neural network. PTEP **2019**(8), 083A04 (2019)
110. Arai, S., Ohzeki, M., Tanaka, K.: Deep neural network detects quantum phase transition. J. Phys. Soc. Jpn. **87**(3), 033001 (2018)
111. Srivastava, R.K., Greff, K., Schmidhuber, J.: Highway networks. arXiv preprint arXiv:1505.00387 (2015)
112. Weinan, E.: A proposal on machine learning via dynamical systems. Commun. Math. Stat. **5**(1), 1–11 (2017)
113. Abarbanel, H.D.I., Rozdeba, P.J., Shirman, S.: Machine learning; deepest learning as statistical data assimilation problems. Neural Comput. **30**(Early Access), 1–31 (2018)
114. Gomez, A.N., Ren, M., Urtasun, R., Grosse, R.B.: The reversible residual network: back propagation without storing activations. In: Advances in Neural Information Processing Systems, pp. 2214–2224 (2017)
115. Haber, E., Ruthotto, L.: Stable architectures for deep neural networks. Inverse Prob. **34**(1), 014004 (2017)
116. Chang, B., Meng, L., Haber, E., Ruthotto, L., Begert, D., Holtham, E.: Reversible architectures for arbitrarily deep residual neural networks. arXiv preprint arXiv:1709.03698 (2017)
117. Chen, T.Q., Rubanova, Y., Bettencourt, J., Duvenaud, D.: Neural ordinary differential equations. arXiv preprint arXiv:1806.07366 (2018)
118. Hashimoto, K., Sugishita, S., Tanaka, A., Tomiya, A.: Deep learning and the AdS/CFT correspondence. Phys. Rev. D **98**(4), 046019 (2018)
119. Lin, H.W., Tegmark, M., Rolnick, D.: Why does deep and cheap learning work so well? J. Stat. Phys. **168**(6), 1223–1247 (2017)
120. Hopfield, J.J.: Neural networks and physical systems with emergent collective computational abilities. Proc. Natl. Acad. Sci. **79**(8), 2554–2558 (1982)

121. Amari, S.-I.: Characteristics of random nets of analog neuron-like elements. IEEE Trans. Syst. Man Cybern. SMC **2**(5), 643–657 (1972)
122. Carleo, G., Troyer, M.: Solving the quantum many-body problem with artificial neural networks. Science **355**(6325), 602–606 (2017)
123. Nomura, Y., Darmawan, A.S., Yamaji, Y., Imada, M.: Restricted boltzmann machine learning for solving strongly correlated quantum systems. Phys. Rev. B **96**(20), 205152 (2017)
124. Carleo, G., Nomura, Y., Imada, M.: Constructing exact representations of quantum many-body systems with deep neural networks. Nat. Commun. **9**(1), 1–11 (2018)
125. Chen, J., Cheng, S., Xie, H., Wang, L., Xiang, T.: Equivalence of restricted boltzmann machines and tensor network states. Phys. Rev. B **97**(8), 085104 (2018)
126. Gao, X., Duan, L.-M.: Efficient representation of quantum many-body states with deep neural networks. Nat. Commun. **8**(1), 662 (2017)
127. Huang, Y., Moore, J.E.: Neural network representation of tensor network and chiral states. arXiv preprint arXiv:1701.06246 (2017)
128. Cohen, N., Shashua, A.: Convolutional rectifier networks as generalized tensor decompositions. In: International Conference on Machine Learning, pp. 955–963 (2016)
129. Hashimoto, K.: D-Brane: Superstrings and New Perspective of Our World. Springer Science & Business Media (2012)
130. Hashimoto, K., Sugishita, S., Tanaka, A., Tomiya, A.: Deep Learning and Holographic QCD. Phys. Rev. D **98**(10), 106014 (2018)
131. He, Y.-H.: Deep-Learning the Landscape. arXiv preprint arXiv:1706.002714 (2017)
132. He, Y.-H.: Machine-learning the string landscape. Phys. Lett. B **774**, 564–568 (2017)
133. Ruehle, F.: Evolving neural networks with genetic algorithms to study the String Landscape. J. High Energy Phys. **8**, 038 (2017)
134. Carifio, J., Halverson, J., Krioukov, D., Nelson, B.D.: Machine learning in the string landscape. J. High Energy Phys. **9**, 157 (2017)
135. Maldacena, J.M.: The Large N limit of superconformal field theories and supergravity. Int. J. Theor. Phys. **38**, 1113–1133 (1999). [Advances in Theoretical and Mathematical Physics **2**, 231 (1998)]
136. Gubser, S.S., Klebanov, I.R., Polyakov, A.M.: Gauge theory correlators from noncritical string theory. Phys. Lett. B **428**, 105–114 (1998)
137. Witten, E.: Anti-de Sitter space and holography. Adv. Theor. Math. Phys. **2**, 253–291 (1998)
138. Susskind, L.: The World as a hologram. J. Math. Phys. **36**, 6377–6396 (1995)
139. Klebanov, I.R., Witten, E.: AdS/CFT correspondence and symmetry breaking. Nucl. Phys. B **556**, 89–114 (1999)
140. Sakai, H., Taguchi, Y., Tokura, Y.: Impact of bicritical fluctuation on magnetocaloric phenomena in perovskite manganites. J. Phys. Soc. Jpn. **78**(11), 113708 (2009)
141. Unger, W.: The chiral phase transition of QCD with 2+1 flavors: a lattice study on goldstone modes and universal scaling. Ph.D. thesis, der Universität Bielefeld (2010)

Index

A. Tanaka et al., *Deep Learning and Physics*, Mathematical Physics Studies,
https://doi.org/10.1007/978-981-33-6108-9

Lightning Source UK Ltd.
Milton Keynes UK
UKHW021254140822
407197UK00006B/22